Standard Model Phenomenology

Standard Model Phenomenology

Shaaban Khalil & Stefano Moretti

CRC Press is an imprint of the
Taylor & Francis Group, an **informa** business

First edition published 2022
by CRC Press
6000 Broken Sound Parkway NW, Suite 300, Boca Raton, FL 33487-2742

and by CRC Press
4 Park Square, Milton Park, Abingdon, Oxon, OX14 4RN

CRC Press is an imprint of Taylor & Francis Group, LLC

ISBN: 978-1-138-33643-8 (hbk)
ISBN: 978-1-032-20168-9 (pbk)
ISBN: 978-0-429-44301-5 (ebk)

DOI: 10.1201/9780429443015

Typeset in LM Roman
by KnowledgeWorks Global Ltd.

Publisher's note: This book has been prepared from camera-ready copy provided by the authors.

To our families

Those near and those far

Contents

Preamble

Surprisingly enough, Science owes to Philosophy the first acceptable definition of the word *phenomenon* in a scientific sense. In fact, it was Immanuel Kant who wrote the following in "The Critique of Pure Reason" [1]: *Phenomena constitute the world as we experience it, as opposed to the world as it exists independently of our experiences* (the latter being indeed made by 'things-in-themselves'). Humans cannot, according to Kant, know things-in-themselves, only things as they experience them. He contrasted the concept of phenomenon against that of *noumenon*, which is the "thing in itself (das Ding an sich)", an allegedly unknowable, undescribable reality that, in some way, lies "behind" observed phenomena. Noumena are sometimes spoken of, though the very notion of accessing items in "the noumenal world" is problematic, since the very notions of number and individuality are among the categories of the understanding, which are supposed to apply only to phenomena, not noumena.

This concept of phenomenon, propagating over centuries from the works of Georg Wilhelm Friedrich Hegel through to those of Martin Heidegger, specifically led to a tradition of philosophy indeed known as 'Phenomenology', as conceived by Edmund Husserl, who established the first school of it. However, in his conception, phenomenology is primarily concerned with the systematic reflection on and study of the structures of consciousness and the phenomena that appear in acts of consciousness. Hence, phenomenology here can be clearly differentiated from the Cartesian method of analysis which sees the world as objects, sets of objects as well as objects acting and reacting upon one another. The latter is what we embrace here.

Phenomenon for us stands for any potentially observable event. Hence, phenomena make up the raw data of science. Wikipedia tells us the following. *In physics, phenomenology is the application of theoretical physics to experimental data by making quantitative predictions based upon known theories. It is in contrast to experimentation in the scientific method, in which the goal of the experiment is to test a scientific hypothesis instead of making predictions.*

Phenomenology is related to the philosophical notion in that these predictions describe anticipated behaviours for the phenomena in reality. We do not disagree with this. In contrast, we disagree with John Archibald Wheeler when he writes: *No phenomenon is a real phenomenon until it is an observed phenomenon.* One should in fact not discount the fact that we may sometimes outsmart Nature, by seeing what She has not yet revealed to us. This is what it means to go beyond. But in order to do so, we must know first what lies before. This is what inspired this book.

Our definition of phenomenology is simple. We apply it to particle physics, but it surpasses it, as it can equally be used in other disciplines. It is nothing but that branch of particle physics that seeks knowledge in a twofold way. On the one hand, by exploiting the hints and clues available in observable phenomena (the aforementioned experimental data), without any preconception on the theory governing the latter. On the other hand, by parametrising theories into a set of predictions, whether observed already or not, that can directly be tested by experiment, thus confirming or disproving the former.

Phenomenology is therefore a bridge between theory and experiment. With this book, we intend to walk this bridge back and forth innumerable times in order to explain to the reader how the SM of particle physics came about. We aim at enthusing him and/or her on what we know, but equally also leave him and/or her with a new appetite, that of going beyond this knowledge. Of building new bridges.

Preface

We have dwelt upon the concept of phenomenon at length in our preamble. However, what is the use of phenomena that we make in Science? We 'reduce' them. Reductionism is the practice of analysing and describing a complex phenomenon in terms of its simple or fundamental constituents, especially when this is deemed to provide a sufficient explanation. The SM is the current frontier of the reductionist approach in particle physics in order to understand matter and forces. This book is dedicated to understanding what the most basic building blocks of matter are and how they interact with each other. But this is just a starting point.

The ultimate goal of such an endeavour is in fact a ToE which mathematically describes these two elements (matter and forces) under all conditions. We might hope on aesthetic grounds for some kind of unification of particle physics leaving a theory with just one species of particle and one species of interaction. Such an idea is strongly hinted at by the progress of particle physics to date, wherein a vast array of phenomena has been distilled into a relatively few laws, *e.g.*, all of the phenomena of electricity and magnetism now reside in a single theory.

However, we cannot, of course, demand this of Nature if She does not want to cooperate. One could imagine a "Russian doll" scenario where matter continues to be found to have smaller and smaller constituents as we look on smaller and smaller scales ad infinitum. Philosophically we may worry, even in the case of a final ToE, why the particular fundamental matter and forces existing within it and the laws governing them are the ones Nature chooses. The answer to such a question, however, seems beyond the power of Science, even in principle. Indeed, the answer to this question appears to belong to the noumenal world. So, we do not bother with it here. In fact, the goal of achieving a complete mathematical description of Nature is per se such a thrilling experience that, as scientists, we are far too taken by understanding how it all came about to worry about asking ourselves why it did.

Indeed, we should finally note that a reductionist approach able to ultimately formulate even a ToE in the simplest possible form does not imply that we will understand everything of the world surrounding us. In fact, for a start, we will not be able to understand why we can explain it. The leap between fundamental particles and interactions and how they combine into producing life, consciousness and ultimately our understanding of their behaviour, no matter how exciting this might be, is far too remote from the scope of particle physics. The latter only deals with the former and so we do here.

This book is based largely on our personal training and research activities with various collaborators, whom we thank here collectively. We cannot in fact acknowledge them one by one, as the list would be far too long. However, we would like to single out here those who can be more closely associated with this book, for one reason or another. We thus thank in particular, in alphabetical order, A. Ballestrero, J.R. Ellis, N.J. Evans, E. Maina, R.G. Roberts, M.H. Seymour, C.H. Shepherd-Themistocleous, P.Z. Skands, W.J. Stirling and B.R. Webber. Finally, in addition to our own institutions, we would like to also thank CERN, where many parts of this book were worked upon. We are also grateful to S. Bradshaw for help with copyright issues.

Shaaban Khalil
Stefano Moretti

Foreword

Particle physics is a journey of exploration deep into the invisible corners of the Universe to uncover the nature of fundamental forces and constituents that hold matter together. This is a journey that started in the early part of the last century, eventually got more sophisticated and led to a remarkable uptake in our knowledge in the last fifty years. As a result, it is now widely believed that, leaving aside a few important puzzles, there is a complete understanding of all observations in Nature. This amazing achievement is based on a framework called the SM invented in the 1960s by Sheldon L. Glashow, Steven Weinberg and Abdus Salam using a deeper understanding of QFT of local symmetries and their breaking.

Despite its successes, the SM is believed to be incomplete due the several puzzles that it does not explain. For example, it does not explain the non-zero neutrino masses established more than two decades ago and does not explain the DM of the universe for which there is now overwhelming evidence. It does not explain the dark energy in the Universe either, that accounts for the bulk of its energy budget. There are theoretical issues as well that the SM faces and that call for its extension.

The journey of exploration therefore goes on and will likely go on for a long time. In fact, the search for BSM physics is almost the exclusive focus of current research by the particle physics community on both the theoretical and experimental fronts. The various boundaries in this exploration are the energy, intensity as well as cosmic frontiers. At the energy frontier, there is the LHC which has been searching in the higher mass domain and proposals for even higher energy colliders using protons and electrons/positrons are now being considered. They will look for higher mass particles as well as long lived ones. They are supplemented by experiments (and plans) at the intensity frontier searching for higher precision effects in observables such as the anomalous magnetic moment of the muon, searches for DM, proton decay and neutron-anti-neutron oscillations, etc. Combine this with the discoveries coming from the cosmic frontier and there opens up a whole list of challenges for the field.

Embarking on this path of discovery in BSM physics, be it experimental or theoretical, requires that one has a thorough grounding in the basics of the SM. That is now the pedestal on which we stand and look up. For that, one needs to be conversant with the founding ideas of QFT of local symmetries which led to the very successful framework of QED and which eventually led to the spontaneously broken gauge theories, the basis of the SM.

This book, 'Standard Model Phenomenology', by two outstanding phenomenologists and model builders, Shaaban Khalil and Stefano Moretti, summarises the necessary background on the SM in a lucid and accessible manner providing a firm stepping stone to do BSM physics. As the authors emphasise, phenomenology builds bridges between theory and experiment allowing experimental tests of the theory and the authors have done an admirable job of it for the SM. The book starts out with a description of the basics of QED followed by the salient features of weak interactions, the $V - A$ theory and how it gets embedded into the gauge theories based on the $SU(3)_C \times SU(2)_L \times U(1)_Y$ local symmetry of the SM. This is followed by an excellent summary of the phenomenological implications of the SM and how it was tested by experiments using high energy hadronic scatterings. Then there is a very useful discussion of the QCD theory of strong interactions, the discovery of the top quark and the ultimate crowning jewel of the SM, the discovery of the Higgs boson in 2012 at the LHC at CERN. Connecting theory to experiment is the main strength of the book. Finally, the book gives a brief introduction to the discoveries involving the neutrino mass and the simple paradigm to understand its smallness.

This book will be an excellent addition to the collection of any particle physics beginner, both experimental and theoretical. I recommend this also as a textbook for any course on the SM.

Rabindra Mohapatra

Acronyms

For the reader's convenience, we introduce here the acronyms that we use in the book. They are listed in alphabetical order. If the same acronym appears more than once, the order here reflects that of their first appearance in the text.

ALEPH	Apparatus for LEP Physics
ATLAS	A Toroidal LHC ApparatuS
BR	Branching Ratio
BSM	Beyond the Standard Model
C	Charge
CERN	Conseil Européen pour la Recherce Nucléaire
CKM	Cabibbo-Kobayashi-Maskawa
CL	Confidence Level
CP	Charge/Parity
CMS	Compact Muon Solenoid
CPT	Chiral Perturbation Theory
CPT	Charge/Parity/Time
DELPHI	Detector with Lepton, Photon and Hadron Identification
DESY	Deutsches Elektronen-Synchrotron

DIS	Deep Inelastic Scattering
DM	Dark Matter
EM	Electro-Magnetic
EW	Electro-Weak
EWSB	EW Symmetry Breaking
FCNC	Flavour Changing Neutral Current
FNAL	Fermi National Accelerator Laboratory
GIM	Glashow-Iliopoulos-Maiani
GUT	Grand Unification Theory
HQET	Heavy Quark Effective Theory
HS	Higgs-Strahlung
IS	Inverse Seesaw
ISR	Initial State Radiation
ISR	Intersecting Storage Rings
L3	3-Layered Detector
LEP	Large Electron Positron
LEPEWWG	LEP Electroweak Working Group
LEPHIGGSWG	LEP Higgs Working Group
LHC	Large Hadron Collider
LO	Leading Order
ME	Matrix Element
NLO	Next-to-LO
NNLO	Next-to-NLO
NNNLO	Next-to-NNLO

NP	New Physics
NuTeV	FNAL Experiment E815
OPAL	Omni-Purpose Apparatus at LEP
OPE	Operator Product Expansion
P	Parity
PDG	Particle Data Group
PMNS	Pontecorvo-Maki-Nakagawa-Sakata
QCD	Quantum Chromo-Dynamic
QED	Quantum Electro-Dynamics
QFT	Quantum Field Theory
RGE	Renormalisation Group Equation
SLAC	Stanford Linear Accelerator Center
SLC	Stanford Linear Collider
SLD	Stanford Linear Detector
SM	Standard Model
SSB	Spontaneous Symmetry Breaking
TGC	Triple Gauge Coupling
ToE	Theory of Everything
UV	Ultra-Violet
$V-A$	Vector minus Axial-Vector
VBF	Vector Boson Fusion
VEV	Vacuum Expectation Value

CHAPTER 1

QED Theory

QED is the fundamental theory that describes the interactions of charged particles with the EM field (light). It is one of the most accurate and best tested theories in the history of Science.

1.1 MAXWELL'S EQUATIONS

Maxwell's equations form the basic of classical EM forces. They describe the electric and magnetic fields generated by distributions of electric charges and currents. They are given by

$$\nabla \times \vec{E} + \frac{\partial \vec{B}}{\partial t} = 0\,, \tag{1.1}$$

$$\nabla \times \vec{B} - \frac{\partial \vec{E}}{\partial t} = \vec{J}\,, \tag{1.2}$$

$$\nabla \cdot \vec{E} = \rho\,, \tag{1.3}$$

$$\nabla \cdot \vec{B} = 0\,, \tag{1.4}$$

where $\vec{E}$ is the electric field, $\vec{B}$ is the magnetic field, ρ is the charge density and $\vec{J}$ is the current density. Recall that the first equation is Faraday's law of induction, the second equation is Ampere's law with Maxwell's modification to include the displacement current $\partial \vec{E}/\partial t$, the third and fourth equations are Gauss' laws for electric and magnetic fields, respectively. One can show that the solutions of these equations can be given, in terms of scalar potential

DOI: 10.1201/9780429443015-1

$\phi(t, \vec{x})$ and vector potential $\vec{A}(t, \vec{x})$, as follows:

$$\vec{B} = \nabla \times \vec{A}, \tag{1.5}$$

$$\vec{E} = -\nabla\phi - \frac{\partial \vec{A}}{\partial t}. \tag{1.6}$$

In field theory, one uses the 4-vector potential $A^\mu = (\phi, \vec{A})$ and defines the EM field tensor as

$$F_{\mu\nu} = \partial_\mu A_\nu - \partial_\nu A_\mu \,, \tag{1.7}$$

which is explicitly given by

$$F^{\mu\nu} = \begin{bmatrix} 0 & -E_x & -E_y & -E_z \\ E_x & 0 & -B_z & B_y \\ E_y & B_z & 0 & -B_x \\ E_z & -B_y & B_x & 0 \end{bmatrix}. \tag{1.8}$$

In terms of $F^{\mu\nu}$, Maxwell's equations can be written as follows:

$$\partial_\mu F^{\mu\nu} = J^\nu \,, \tag{1.9}$$

$$\partial^\alpha F^{\mu\nu} + \partial^\beta F^{\alpha\mu} + \partial^\mu F^{\nu\alpha} = 0. \tag{1.10}$$

Here, a few remarks are in order. (i) Maxwell's equations clearly confirmed the unification between electric and magnetic forces and explicitly showed that they can be expressed in terms of the vector field A_μ; therefore, the EM theory is considered the first successful (classical) unified field theory. (ii) The electric and magnetic fields are determined through Maxwell's equations in terms of electric charges (but no magnetic charges, hence, no monopoles exist). (iii) These fields exert influences on the charges (matter fields) via the Lorentz force, which is given by

$$\vec{F} = q\vec{E} + \vec{v} \times \vec{B}, \tag{1.11}$$

where $\vec{v}$ is the velocity of a charged particle.

1.2 $U(1)$ ABELIAN GAUGE SYMMETRY

The theory of EM fields has the structure of an Abelian gauge field theory. It can be derived from the theory of a free electron by requiring it to be gauge invariant. Consider the Lagrangian (density) of a free electron field $\psi(x)$,

$$\mathscr{L}_0(\psi, \bar{\psi}) = \bar{\psi}(x)(i\gamma^\mu \partial_\mu - m)\psi(x), \tag{1.12}$$

where the γ^μ's are the Dirac matrices, which are given as $(i = 1, ..., 3)$, *e.g.*,

$$\gamma^0 = \begin{pmatrix} 0 & 1 \\ 1 & 0 \end{pmatrix}, \quad \gamma^i = \begin{pmatrix} 0 & \sigma^i \\ -\sigma^i & 0 \end{pmatrix}, \tag{1.13}$$

in terms of the σ^i's, which are the Pauli matrices. Also, $\bar{\psi}(x)$ is defined as $\bar{\psi}(x) = \psi^\dagger(x)\gamma^0$. Clearly the Lagrangian $\mathscr{L}_0(\psi, \bar{\psi})$ is invariant under the *global* transformation

$$\psi'(x) = e^{-i\alpha}\psi(x)\ , \quad \alpha = \text{constant}. \tag{1.14}$$

Invariance of the Lagrangian under the transformation in Eq. (1.14) means that the addition of any constant value to the phase of the function $\psi(x)$ does not lead to any physical consequences.

Now, we turn this symmetry into a *local* gauge symmetry, *i.e.*, by replacing the constant α with an arbitrary real function of the space-time points, $\alpha(x)$:

$$\psi'(x) = e^{-i\alpha(x)}\ \psi(x). \tag{1.15}$$

Here, the gauge symmetry group is $U(1)$; hence it is referred to as an Abelian gauge invariance. However, one can easily observe that the Lagrangian in Eq. (1.12) is no longer invariant under this local transformation. In fact, under the latter, one finds

$$\mathscr{L}_0 \to \mathscr{L}_0 + \bar{\psi}\gamma^\mu\psi\ \partial_\mu\alpha(x). \tag{1.16}$$

Invariance under the local gauge transformation in Eq. (1.16) can be obtained if the electron field couples with a vector gauge field A_μ. This can be ensured by replacing the normal derivative ∂_μ by the covariant derivative D_μ, defined by

$$D_\mu\psi = (\partial_\mu + ieA_\mu)\psi. \tag{1.17}$$

Also, the vector field should transform under Eq. (1.16) as follows:

$$A'_\mu = A_\mu + \frac{1}{e}\partial_\mu \alpha(x). \tag{1.18}$$

In this case, $D_\mu \psi$ transforms under the gauge transformation exactly as ψ:

$$D_\mu \psi \to e^{-i\alpha(x)} D_\mu \psi. \tag{1.19}$$

Therefore, the following Lagrangian is gauge invariant:

$$\begin{aligned} \mathscr{L}_1 &= \bar{\psi}\left(i\gamma^\mu D_\mu - m\right)\psi \\ &= \mathscr{L}_0 + \mathscr{L}_I, \end{aligned} \tag{1.20}$$

where $\mathscr{L}_I$ is the interaction term between the EM field and the electron. It is given by

$$\mathscr{L}_I = -e\bar{\psi}\gamma^\mu\psi\, A_\mu. \tag{1.21}$$

The free Lagrangian of the EM field that describes the dynamics of the gauge field A_μ should be included into the total Lagrangian of the system. The field strength tensor

$$F_{\mu\nu} = \partial_\mu A_\nu - \partial_\nu A_\mu \tag{1.22}$$

is invariant under the gauge transformation in Eq. (1.18). Therefore, the simplest invariant Lagrangian for a free EM field is proportional to the scalar quantity $F^{\mu\nu}F_{\mu\nu}$, *i.e.*,

$$\mathscr{L}_A = -\frac{1}{4}F_{\mu\nu}F^{\mu\nu}. \tag{1.23}$$

Thus one can write the final gauge invariant Lagrangian of QED as

$$\mathscr{L} = \bar{\psi}\left(i\gamma^\mu\partial_\mu - m\right)\psi - e\bar{\psi}\gamma^\mu\psi A_\mu - \frac{1}{4}F_{\mu\nu}F^{\mu\nu}, \tag{1.24}$$

where the factor $-1/4$ has been introduced for convenience. The equation of motion for the field A_μ is

$$\partial_\mu F^{\mu\nu} = eJ^\mu, \tag{1.25}$$

where $J^\mu \equiv \bar{\psi}\gamma^\mu\psi$ is the Noether current generated by the $U(1)$ symmetry. One can show that this equation is equivalent to Gauss' and Ampère's laws in the 4-vector representation. A particular choice of A^μ restricts it to obey the Jacobi' identity

$$[D_\mu, [D_\nu, D_\lambda]] + [D_\lambda, [D_\mu, D_\nu]] + [D_\nu, [D_\lambda, D_\mu]] = 0. \tag{1.26}$$

This constraint in turn imposes the field strength to satisfy the Bianchi identity:

$$\partial_\mu F_{\nu\lambda} + \partial_\lambda F_{\mu\nu} + \partial_\nu F_{\lambda\nu} = 0. \tag{1.27}$$

The explicit form of this equation verifies that it is equivalent to Faraday's law of induction and Gauss' law for magnetism.

Two remarks about the QED Lagrangian in Eq. (1.24) are in order: (i) the photon mass term $\frac{1}{2}m^2 A^\mu A_\mu$ is forbidden by gauge symmetry, since it is not invariant under the gauge transformation of Eq. (1.18); (ii) there is no gauge field self-coupling, because the field A_μ does not carry a charge (*i.e.*, a $U(1)$ quantum number). Therefore, without matter the EM theory is a free field theory.

In vacuum, Maxwell's equations can be written as

$$\partial_\mu F^{\mu\nu} = \partial_\mu(\partial^\mu A^\nu - \partial^\nu A^\mu) = 0. \tag{1.28}$$

From the gauge transformation in Eq. (1.18), one can choose $\alpha(x)$ such that $\Box\,\alpha(x) = -\partial_\mu A^\mu(x)$, where $\Box = \partial_\mu \partial^\mu$. In this case, one gets

$$\partial_\mu A^\mu = 0, \tag{1.29}$$

which is called the Lorenz gauge. This condition effectively reduces the number of independent components of A^μ from four to three. However, the Lorenz gauge does not make A^μ unique. It is clear that if A^μ satisfies the Lorentz condition, one can choose the function $\alpha(x)$ such that (see Eq. (1.18))

$$\frac{\partial \alpha}{\partial t} = -eA_0 \Longrightarrow A_0' = 0. \tag{1.30}$$

In this case, the Lorenz gauge implies that $\nabla \cdot A' = 0$. Potentials satisfying the conditions

$$A_0 = 0 \;\text{ and }\; \nabla \cdot A = 0 \tag{1.31}$$

are said to belong to the radiation (or Coulomb) gauge. In this gauge there are only two independent components of A^μ, the two transverse components of the polarisation vector. This is the case in the real world, so the physical nature of the EM field is consistent with the radiation gauge.

1.3 QUANTISATION OF THE EM FIELD

Here we briefly review the quantisation of the EM field. We start with the canonical quantisation where the 4-vector field $A_\mu(x)$ is treated as operator and time-like commutation relations between A_μ and its conjugate momentum are imposed. The conjugate momentum associated with the field $A_\mu(x)$ is defined by

$$\pi^\mu = \frac{\partial \mathcal{L}}{\partial \dot{A}_\mu} = -F^{0\mu}. \tag{1.32}$$

This means that π^0 is identically zero and $A_0(x)$ is a classical function (*i.e.*, it commutes with all field variables) and can be set to zero, *i.e.*, $A_0(x) = 0$. This is a consequence of the gauge invariance of the EM theory. The above choice is known as a temporal gauge. In this gauge one finds that the non-vanishing commutation relation is given by

$$[A_i(x), \pi^j(y)]|_{x^0=y^0} = [A_i(x), E^j(y)]|_{x^0=y^0} = i\delta_i^j \delta^3(x-y), \tag{1.33}$$

where the conjugate momentum π^j is defined as

$$\pi^i(x) = -F^{0i} = E^i = -\dot{A}^i(x). \tag{1.34}$$

As mentioned, from the Lorenz gauge and with the choice $A_0 = 0$, one obtains the condition

$$\nabla \cdot \mathbf{A} = 0, \tag{1.35}$$

so that, in this gauge, there are only two independent transverse components of A^μ. Thus, it is clear that the commutation relation in Eq. (1.33) is inconsistent with the transversality condition in Eq. (1.35) and must be modified. The proper commutation relation is given by

$$[A_i(x), E^j(y)]_{x^0=y^0} = i\delta_{i\mathrm{TR}}^j(x-y), \tag{1.36}$$

where $\delta^j_{i\text{TR}}(x-y)$ is defined as

$$\delta^j_{i\text{TR}}(x-y) = \left(\delta^j_i - \frac{\partial_i \partial^j}{\nabla^2}\right)\delta^3(x-y). \tag{1.37}$$

Hence, $\partial^i \delta^j_{i\text{TR}}(x-y) = 0$ and the new commutation relation in Eq. (1.36) is consistent with the transverse condition in Eq. (1.35). In addition, we have the usual commutation relations:

$$[A_i(x), A_j(y)]_{x^0=y^0} = [E^i(x), E^j(y)]_{x^0=y^0} = 0. \tag{1.38}$$

Now one can define the Hamiltonian density of the EM field theory as

$$\mathcal{H} = \pi^i \dot{A}_i - \mathcal{L}. \tag{1.39}$$

Using the relations $\pi^i = -F^{0i} = E^i$ and $F^{ij} = \epsilon^{ijk} B_k$, one gets the usual expression of energy in classical electro-dynamics:

$$H = \int d^3x \mathcal{H} = \frac{1}{2}\int d^3x \ (\mathbf{E}^2 + \mathbf{B}^2). \tag{1.40}$$

In the temporal gauge $A_0 = 0$ and with the condition $\nabla \cdot \mathbf{A} = 0$, the wave equation in Eq. (1.28) takes the form:

$$\Box A^i(x) = 0. \tag{1.41}$$

The general solution of this field equation can be expanded in terms of plane waves as follows:

$$\mathbf{A}(x) = \int \frac{d^3k}{\sqrt{(2\pi)^3 2k_0}} \sum_{\lambda=1}^{2} \epsilon^\lambda(k)\left[a^{(\lambda)}(k)e^{-ikx} + a^{(\lambda)^+}(k)e^{ikx}\right], \tag{1.42}$$

with $k^2 = 0$, *i.e.*, $k_0 = |\mathbf{k}|$. Note that $a^{(\lambda)}(k)$ and $a^{(\lambda)^+}(k)$ are operators and the field $\mathbf{A}(x)$ is defined as an Hermitian operator. The condition $\nabla \cdot \mathbf{A} = 0$ implies that

$$\mathbf{k} \cdot \epsilon^\lambda(k) = 0. \tag{1.43}$$

Since e^{ikx} and e^{-ikx} are orthogonal functions and ϵ^λ and $\epsilon^{\lambda'}$ are orthonormal vectors, *i.e.*, $\epsilon^\lambda \cdot \epsilon^{\lambda'} = \delta^{\lambda\lambda'}$, one can show that

$$a^{(\lambda)}(k) = i \int \frac{d^3k}{\sqrt{(2\pi)^3 2k_0}} \left(e^{ikx} \overleftrightarrow{\partial}_0 \mathbf{A}(x)\right) \cdot \epsilon^\lambda(k), \tag{1.44}$$

$$a^{(\lambda)^+}(k) = -i \int \frac{d^3k}{\sqrt{(2\pi)^3 2k_0}} \left(e^{-ikx} \overleftrightarrow{\partial}_0 \mathbf{A}(x)\right) \cdot \epsilon^\lambda(k). \tag{1.45}$$

Using the commutation relations in Eqs. (1.36) and (1.38), one can show that

$$[a^{(\lambda)}(k), a^{(\lambda')^+}(k')] = \delta_{\lambda\lambda'}\delta^3(k-k'), \tag{1.46}$$

$$[a^{(\lambda)}(k), a^{(\lambda')}(k')] = [a^{(\lambda)^+}(k), a^{(\lambda')}(k')] = 0. \tag{1.47}$$

Therefore, the operators $a^{(\lambda)}(k)$ and $a^{(\lambda)^+}(k)$ are interpreted as annihilation and creation operators for photons. The Hamiltonian of the EM field can then be expressed in terms of the annihilation and the creation operators. After some algebra, one finds that

$$H = \sum_{\lambda=1}^{2} \int d^3k \frac{k_0}{2} \left[a^{(\lambda)}(k)\, a^{(\lambda)^+}(k) + a^{(\lambda)^+}(k)\, a^{(\lambda)}(k)\right]. \tag{1.48}$$

Using normal ordering to remove the vacuum energy, we have

$$H = \sum_{\lambda=1}^{2} \int d^3k k_0\, a^{(\lambda)}(k)\, a^{(\lambda)^+}(k). \tag{1.49}$$

The Coulomb gauge has the advantage that the physical degrees of freedom are manifest. However, in this gauge the Lorentz invariance is lost.

We turn now to the explicit quantisation of the EM field through the path integral formalism, which manifests Lorentz covariance. The partition function for QED is given by

$$Z[J] = \int DAe^{iS[A,J]}, \tag{1.50}$$

where $S[A,J]$ is defined in momentum space as

$$S[A,J] = \frac{1}{2}\frac{d^4k}{(2\pi)^4} \times \left[-k^2 \tilde{A}_\mu \left(\eta_{\mu\nu} - \frac{k^\mu k^\nu}{k^2}\right) \tilde{A}_\nu(-k) + \tilde{J}^\mu(k)\tilde{A}_\mu(-k) + \tilde{J}^\mu(-k)\tilde{A}_\mu(k)\right]. \tag{1.51}$$

However, due to the gauge symmetry and the fact that two gauge equivalent configurations A_μ and A'_μ would yield the same $S[A, J]$, a redundant integration over a continuous infinity of physically equivalent field configurations is performed that would make the integral in the definition of partition function $Z[j]$ diverge. This divergence is of course non-physical. This relates to the problem that the matrix $(\eta_{\mu\nu} - \frac{k^\mu k^\nu}{k^2})$, which inverse defines the propagator of the photon, is singular.

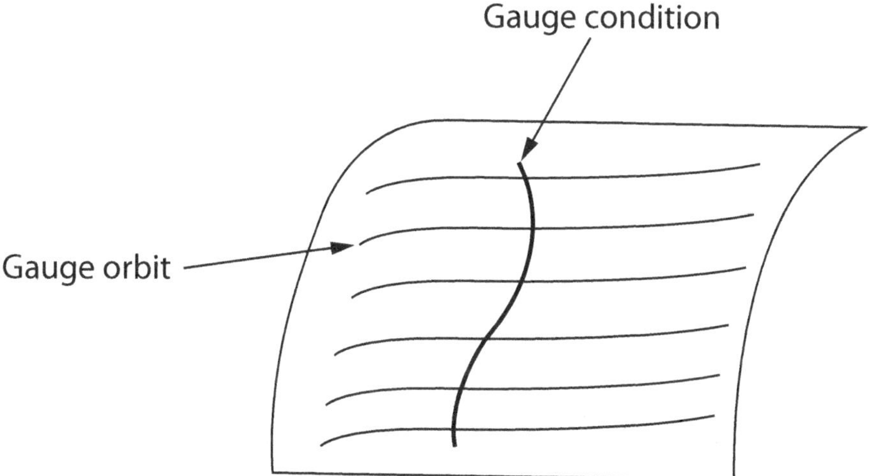

Figure 1.1 Schematic representation of the gauge fixing condition.

To avoid this problem of overcounting, one must select only one physical equivalent field $A^\alpha_\mu(x)$, which obeys some gauge-fixing condition $g(A^\alpha_\mu(x)) = 0$, as shown in Fig. 1.1. The physics must be gauge invariant, so it remains independent from the choice of the gauge fixing. This operation is called the Faddeev-Popov prescription. In this method, one begins with the following integral

$$\Delta_g^{-1}[A_\mu] = \int D\alpha\ \delta(g(A^\alpha_\mu(x)). \tag{1.52}$$

One can easily show that $\Delta_g^{-1}[A_\mu]$ is gauge invariant. Also, the above equation can be written as

$$1 = \Delta_g[A_\mu] \int D\alpha\ \delta(g(A^\alpha_\mu(x)), \tag{1.53}$$

which can be inserted in the definition of $Z[J]$ as follows:

$$Z[J] = \left[\int D\alpha\right] \int DA\ \Delta_g[A_\mu]\ \delta(g(A_\mu))\ e^{iS[A,J]}. \tag{1.54}$$

The factor in brackets in the above equation is an infinite constant, which is a result of summing over gauge equivalent states. For infinitesimal $\alpha(x)$, we can write $g(A_\mu^\alpha)$ as

$$g(A_\mu^\alpha) = \frac{\partial g}{\partial A_\mu}\partial_\mu \alpha. \tag{1.55}$$

Thus

$$\Delta(A)^{-1} = \int D\alpha \delta\left(\frac{\partial g}{\partial A_\mu}\partial_\mu \alpha\right) = \left|\frac{\partial g}{\partial A_\mu}\partial_\mu\right|^{-1}. \tag{1.56}$$

Here we have used the identity $\delta(Mc) = \frac{1}{\det M}\delta(c)$, with $M_b^a(x,x') = \frac{\partial g^a(A(x))}{\partial A_\mu^b(x')}\partial_\mu$ (for a general gauge symmetry group). One can write the above determinant, by introducing two anti-commuting fields η^a and $\bar{\eta}^a$, in the following form:

$$\Delta(A) = \int D\bar{\eta} D\eta \, e^{\frac{i}{\hbar}\int d^4x d^4x' \bar{\eta}^b(x') M_b^a(x,x') \eta^a(x)}. \tag{1.57}$$

As an example, we consider the Lorenz gauge condition, *i.e.*, $g(A) = \partial^\mu A_\mu$. In this case, $M(x,x') = \partial^2 \delta(x-x')$ and the action of η and $\bar{\eta}$ becomes $\int d^4x \, \partial_\mu \bar{\eta} \, \partial^\mu \eta$.

One may write $\delta(g(A)) = \int D\chi e^{i\chi \, g(A)}$, then the generating function Z is given by

$$Z[J] = \mathcal{N} \int DA \; D\bar{\eta} \; D\eta D\chi e^{iS_{\text{tot}}}, \tag{1.58}$$

where S_{tot} is given by

$$S_{\text{tot}} = S[A] + \int d^4x \left(J.A + \chi g(A) + \int d^4x' \bar{\eta}(x') M(x,x') \eta(x) \right). \tag{1.59}$$

To get Feynman rules we follow the usual procedure of writing the interacting Lagrangian as a function of functional derivatives. After adding the gauge fixing term to the Lagrangian, in the form $-\frac{(\partial_\mu A^\mu)^2}{2\xi}$, we have the following generating functional:

$$\int DA \; e^{\int d^4x \left[\frac{i}{2} A_\mu(x)(\Box \eta^{\mu\nu} - \partial^\mu \partial^\nu) A_\nu(x) + J_\mu A^\mu\right]} = \mathcal{N} e^{\int d^4x d^4y J_\mu(x)\left(\Box \eta^{\mu\nu} - \partial^\mu \partial^\nu (1-\frac{1}{\xi})\right) J_\nu(y)}, \tag{1.60}$$

which leads to the following photon propagator and the three point interaction:

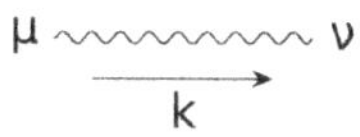

$$\mu \; \nu \quad k \qquad \frac{-i}{k^2+i\varepsilon}\ \eta^{\mu\nu} - (1-\xi)\frac{k^\mu k^\nu}{k^2}\ .$$

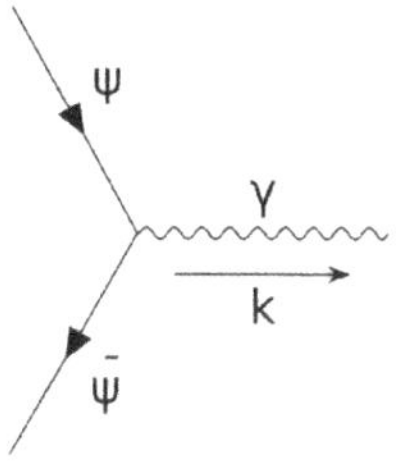

$$\psi \quad \gamma \quad k \quad \bar{\psi} \qquad = ie\gamma^\mu$$

(Other Feynman rules of the SM can be found in Appendix A.)

1.4 QED AND THE ELECTRON ANOMALOUS MAGNETIC MOMENT

The prediction of the anomalous magnetic moment of the electron is one of the stunning achievements of QED. Nowadays, the theoretical prediction agrees with experimental measurements up to one part in a trillion.

The magnetic moment $\vec{\mu}$ of the electron is defined in terms of its spin $\vec{s}$ as

$$\vec{\mu} = g_e \frac{e\hbar}{2m_e c}\vec{s}, \tag{1.61}$$

where g_e is the gyromagnetic ratio of the electron which, according to the Dirac equation, is given by $g_e = 2$. However, experimental measurements confirmed that $g_e \neq 2$. Therefore, one defines the anomalous magnetic moment of the electron, a_e, as

$$a_e = \frac{g_e - 2}{2}. \tag{1.62}$$

Today, the experimental results of a_e [2] is given by

$$a_e^{\rm exp} = 15965218873(28) \times 10^{-14}. \tag{1.63}$$

In QED, a non-vanishing value of a_e can be obtained through radiative (photon and electron) loop corrections. These corrections can be calculated

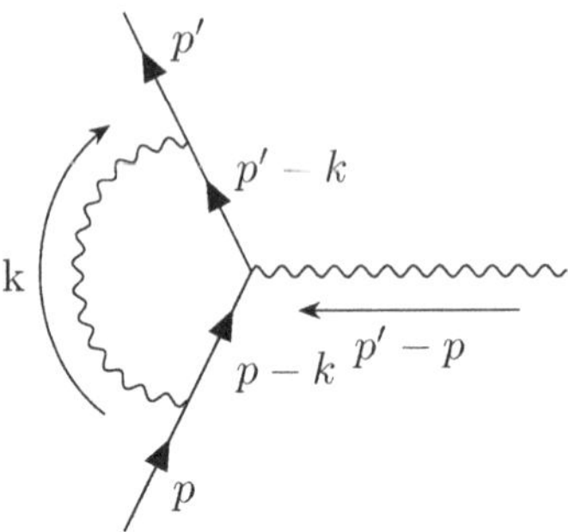

Figure 1.2 Second order vertex correction.

by considering the type of interactions shown in Fig. 1.2 and given by

$$\mathscr{L}_{\text{int}} = -e\bar{\ell}(p')\gamma^{\mu}\ell(p)A_{\mu}(q), \tag{1.64}$$

where $q = p' - p$ and $\ell(p)$ is the lepton wave function of momentum p. The radiative correction to the scattering amplitude in momentum space can be written as

$$i\mathcal{M} = -ie\bar{u}(p')\ \Gamma^{\mu}\ u(p)\ A_{\mu}(q), \tag{1.65}$$

where the effective coupling Γ^{μ}, at one loop, is given by

$$\Gamma^{\mu} = -ie^2 \int \frac{d^4k}{(2\pi)^4} \frac{\gamma^{\nu}(\not{p}' - \not{k} + m)\gamma_{\mu}(\not{p} - \not{k} + m)\gamma_{\nu}}{[(p'-k)^2 - m^2][(p-k)^2 - m^2]k^2}. \tag{1.66}$$

Using Feynman parameterisation to compute the above integral,

$$\frac{1}{a_1 a_2 a_3} = 2\int_0^1 dx \int_0^{1-x} dy \frac{1}{(xa_1 + ya_2 + (1-x-y)a_3)^3}, \tag{1.67}$$

one finds

$$\Gamma^{\mu} = -ie^2 4m(p^{\mu} + p'^{\mu}) \int_0^1 dx \int_0^{1-x} dy \int \frac{d^4\ell}{(2\pi)^4} \frac{(x+y)(1-(x+y))}{[\ell^2 - m^2(x+y)^2]^3}, \tag{1.68}$$

where $\ell = k + (xp' + yp)$. So Γ^{μ} becomes

$$\Gamma^{\mu} = -\frac{e^2}{16\pi^2 m}(p^{\mu} + p'^{\mu}) = -\frac{\alpha}{4\pi m}(p^{\mu} + p'^{\mu}), \tag{1.69}$$

where α is the fine structure constant. From Gordon decomposition,

$$\bar{u}\gamma^\mu u = \bar{u}\left[\frac{(p+p')^\mu}{2m} + \frac{i}{2m}\sigma^{\mu\nu}(p'-p)_\nu\right]u, \tag{1.70}$$

one can write Γ^μ in the following general form:

$$\Gamma^\mu(p,q) = F_1(q^2)\gamma^\mu + F_2(q^2)\frac{i\sigma^{\mu\nu}}{2m}q_\nu. \tag{1.71}$$

The contribution from F_1 is deduced only at the tree level and all corrections to $F_1(0)$ vanish, by the field strength renormalisation. Furthermore, $F_2(0) = 0$ at tree level and beyond LO corresponds to the anomalous magnetic moment, namely

$$a_e = \frac{g-2}{2} = F_2(0). \tag{1.72}$$

From our explicit calculation, one finds that the anomalous magnetic moment of the electron at one loop is given by $\alpha/2\pi$. In general, the effective perturbation series for a QED calculation can be expressed as powers of (α/π):

$$a_e(\text{QED}) = C_1\left(\frac{\alpha}{\pi}\right) + C_2\left(\frac{\alpha}{\pi}\right)^2 + C_3\left(\frac{\alpha}{\pi}\right)^3 + C_4\left(\frac{\alpha}{\pi}\right)^4 + C_5\left(\frac{\alpha}{\pi}\right)^5 + \ldots, \tag{1.73}$$

where $C_1 = 1/2$, $C_2 = -0.328478965579$, $C_3 = 1.181241456$, $C_4 = -1.9144(35)$ and $C_5 = 0.0 \pm 4.0$. Thus, one finds [3, 4]

$$a_e^{\text{QED}} = 115965218164(76) \times 10^{-14}, \tag{1.74}$$

which is consistent with the experimental result in Eq. (1.63) up to more than 10 significant figures, which is one of the best agreements between theoretical predictions and experimental measurements known in the history of physics.

CHAPTER 2

From Fermi Theory to $V - A$

The first theory of weak interactions was developed by Enrico Fermi in 1933 [5] to explain the β-decay process

$$n \to p + e + \bar{\nu}, \tag{2.1}$$

where $\bar{\nu}$ is an antineutrino, a particle postulated by Wolfgang Pauli in 1930 in order to preserve energy-momentum as well as spin conservation. Fermi took as a basis the assumption that the nucleus consists of protons and neutrons. He assumed that an electron-neutrino pair is emitted when a neutron is transformed into a proton analogously to when a γ-ray is emitted if an electron in an atom undergoes a transition from one state to another.

2.1 FERMI THEORY

As explained in the previous chapter, the interaction between an electron and the radiation field in QED in described by the "minimal" Lagrangian density in Eq. (1.21)

$$\mathscr{L}_I^{\mathrm{EM}} = -e\bar{\psi}_e\gamma_\mu\psi_e A^\mu, \tag{2.2}$$

where A^μ is the 4-vector potential of the radiation field and $J_\mu \equiv \bar{\psi}_e\gamma_\mu\psi_e$ is the current density associated with the electron field ψ_e, with e being the charge of the electron. Fermi proposed a similar treatment for β-decay by assigning

DOI: 10.1201/9780429443015-2

the role of a photon to the $e\bar{\nu}$ pair suggesting the following replacements

$$\bar{\psi}_e\gamma_\mu\psi_e \to \bar{\psi}_p\gamma_\mu\psi_n, \tag{2.3}$$

$$A^\mu \to \bar{\psi}_e\gamma_\mu\psi_\nu. \tag{2.4}$$

It was taken for granted that this new force should obey P conservation just like the EM force. Thus, the Lorentz-invariant Hamiltonian would be given by

$$\mathscr{H}^\beta = G_F(\bar{\psi}_p\gamma_\mu\psi_n)(\bar{\psi}_e\gamma_\mu\psi_\nu), \tag{2.5}$$

where G_F is the interaction constant (the Fermi coupling) which in the natural unit system has the dimension of inverse mass squared (*i.e.*, $[G_F] = M^{-2}$) while the EM coupling in Eq. (2.2) is dimensionless ($[e] = 1$). This difference indicates the fundamental difference between the fermion-boson interaction described by Eq. (2.2) and the 4-fermion interaction described by Eq. (2.5). This difference will play an important role in the subsequent discussion on the renormalisability of weak interactions.

The interaction as discussed thus far, however, is not Hermitian. Since the Lagrangian must be Hermitian, the conjugate term should be added in the treatment. In 1934, the first positive β^+-decays were observed by Frédéric Joliot and Irène Curie [6] and, a year later, they also discovered the electron capture process $e + p \to n + \nu$, successfully described by the Hamiltonian which is simply conjugate to Eq. (2.5), *i.e.*:

$$\mathscr{H}^{\beta\prime} = G(\bar{\psi}_n\gamma_\mu\psi_p)(\bar{\psi}_\nu\gamma_\mu\psi_e). \tag{2.6}$$

The ensuing Fermi theory was very accurate in calculating transition probabilities in the first order of perturbation and could account for many observed characteristic of β-decay. Hence, the 4-point Fermi coupling was not questioned until difficulties arose in the K^+-meson decay (known as $\tau - \theta$ puzzle) and Wu's experiment (to be discussed in the next section), which confirmed that P is violated in the weak interaction.

It is worth noting that each of the four particles in the β-decay process described by Eq. (2.1) can be represented by a 4-component wavefunction that contributes to the Hamiltonian describing the process. Fortunately, we do not have to study the 256 possible scalars constructed from the four 4-component wavefunctions. Assuming invariance under Lorentz transformations, we are only left with the following five couplings constructed from bilinear covariants:

$$(\bar{\psi}_p \psi_n)\ (\bar{\psi}_e \psi_\nu) \qquad \text{Scalar} \times \text{Scalar}\ \ (S), \tag{2.7}$$
$$(\bar{\psi}_p \gamma_\alpha \psi_n)\ (\bar{\psi}_e \gamma^\alpha \psi_\nu) \qquad \text{Vector} \times \text{Vector}\ \ (V), \tag{2.8}$$
$$(\bar{\psi}_p \sigma_{\alpha\beta} \psi_n)\ (\bar{\psi}_e \sigma^{\alpha\beta} \psi_\nu) \qquad \text{Tensor} \times \text{Tensor}\ \ (T), \tag{2.9}$$
$$(\bar{\psi}_p \gamma_\alpha \gamma_5 \psi_n)\ (\bar{\psi}_e \gamma^\alpha \gamma_5 \psi_\nu) \qquad \text{Axial Vector} \times \text{Axial Vector}\ \ (A), \tag{2.10}$$
$$(\bar{\psi}_p \gamma_5 \psi_n)\ (\bar{\psi}_e \gamma_5 \psi_\nu) \qquad \text{Pseudoscalar} \times \text{Pseudoscalar}\ \ (P). \tag{2.11}$$

Therefore, the most general Lorentz and P invariant Hamiltonian takes the following form:

$$\mathscr{H}^\beta = \sum_i G_i (\bar{\psi}_p \Gamma_i \psi_n)(\bar{\psi}_e \Gamma_i \psi_\nu) + h.c., \tag{2.12}$$

where the matrices Γ^i are, respectively to S, V, T, A and P, given by

$$\Gamma_i = \{I, \gamma^\mu, \sigma^{\mu\nu}, \gamma^\mu \gamma^5, \gamma^5\}, \tag{2.13}$$

with

$$\sigma_{\alpha\beta} \equiv \frac{1}{2i}(\gamma_\alpha \gamma_\beta - \gamma_\beta \gamma_\alpha). \tag{2.14}$$

(Herein, the G_i's are different but each is proportional to G_F.) However, the data of β-decay spectra could be described assuming that only the S and T terms or the V and A ones contribute to the effective Hamiltonian in Eq. (2.12).

2.2 PARITY CONSERVATION IN DOUBT

P is the conserved quantity of space inversion symmetry, $(x, y, z) \to (x', y', z') = (-x, -y, -z)$. The QED Lagrangian is de facto invariant under this symmetry (we use here the shorthand notation x' for all spetial coordinates)

$$\begin{aligned} \mathscr{L}'(x') &= \bar{\psi}'(x')[i\gamma_0 \partial_t + i\gamma \cdot \partial_i' - m]\psi'(x') \\ &= \bar{\psi}(x)\gamma_0[i\gamma_0 \partial_t - i\gamma \cdot \partial_i - m]\gamma_0 \psi(x) \\ &= \bar{\psi}(x)\gamma_0\gamma_0[i\gamma_0 \partial_t + i\gamma \cdot \partial_i - m]\gamma_0 \psi(x) i = \mathscr{L}(x). \end{aligned} \tag{2.15}$$

For many years, invariance under reflection symmetry in fundamental processes (*i.e.*, P invariance) was an act of faith for physicists until doubts were raised by the $\theta - \tau$ puzzle and confirmed by Wu's experiment [7]. In 1957 Pauli wrote about this experiment: *I don't believe that the Lord is a weak left-hander and I am ready to bet a very high sum that the experiment will give a symmetric angular distribution of electrons. I do not see any logical connection between the strength of an interaction and its mirror invariance.* Luckily for him, we are unaware of any bets been taken!

2.2.1 The $\theta - \tau$ Puzzle

The following two processes were observed experimentally:

$$\theta^+ \to \pi^+ + \pi^0, \tag{2.16}$$

$$\tau^+ \to \pi^+ + \pi^+ + \pi^-, \tag{2.17}$$

wherein θ^+ and τ^+ were regarded as two different particles. This can be understood as follows: if P is conserved then $P(\theta^+) = (P(\pi))^2 = (-1)^2 = 1$ and $P(\tau^+) = (P(\pi))^3 = (-1)^3 = -1$. However, experiments indicated that their mass, charge, spin and lifetime are almost identical. This puzzle was the major discussion at the Sixth Rochester Conference on High Energy Nuclear Physics at Rochester, New York, in April 1956. Lee and Yang [8] proposed the hypothesis that θ^+ and τ^+ are the same particle K^+ but could not reach a definite conclusion about conservation of P in weak interaction. Block convinced Feynman to bring up the idea that P in not conserved as people would not take it seriously if he himself brought it up. So Feynman did say during the conference: *Could it be that the θ and τ are different P states of the same particle which has no definite P, i.e., that P is not conserved. That is, does Nature have a way of defining right or left-handedness uniquely? Yang stated that he and Lee looked into that matter without arriving at any definite conclusion.* "Alea iacta est" ("The die has been cast"), one may well have said.

In June 1956, Lee and Yang published their findings and proposed experiments to measure quantities that should not have preferred direction if P is conserved [8], among which was the experiment to measure the projection of the electron momentum on the spin of the cobalt nucleus in a ^{60}Co β-decay process.

2.2.2 Wu's Experiment

On December 1956, after months of struggling with technical difficulties, Madame Wu conducted the proposed experiment to observe the following β-decay process [7]

$$^{60}\text{Co} \rightarrow ^{60}\text{Ni} + e + \bar{\nu_e} + 2\gamma. \tag{2.18}$$

The Cobalt has a spin of $5\hbar$ and Nickel has a spin of $4\hbar$; hence the outgoing electron and antineutrino would have a spin of $+\frac{1\hbar}{2}$ in the same direction as that of Cobalt. By aligning the spin of the Cobalt in the $+z$ direction under very low temperature, we can predict the direction of spin for the outgoing particles. If we arrange a mirror image of this setup, P conservation can be studied by looking at the momentum of the outgoing electrons. Spin will be unchanged in the mirror because it is an axial vector. If P holds, the number of electrons coming out at angle θ should be equal to that at angle $\pi - \theta$ in the mirror.

The general expression for the probability of nuclei with polarised spin $\vec{J}$ to emit electrons with momentum $\vec{P}$ is given by

$$\mathcal{W}_J(P) = \mathcal{W}_0(1 + \alpha\vec{J}.\vec{P}), \tag{2.19}$$

where $\mathcal{W}_0$ is the probability of unpolarised nuclei to emit electrons whereas the parameter α characterises the asymmetry in the emission of electrons with respect to the direction of the polarisation vector of a nucleus. If P is conserved in the decay, then it is clear that

$$\mathcal{W}_J(P) = \mathcal{W}_J(-P). \tag{2.20}$$

Thus $\alpha = 0$. However, Wu showed that $\alpha \simeq -0.7$, which proves that the weak interaction responsible for the β-decay does not conserve P. The results confirmed that electrons are much more likely to emerge at an angle π with respect to their spin. This indicates that electrons have a negative helicity, $h \equiv s \cdot P$, *i.e.*, electrons are left-handed.

In 1957, Garwin, Lederman and Weinrich conducted another experiment [9] that was also suggested by Lee and Yang in their paper using Columbia's

385-MeV synchrocyclotron to observe the $\pi - \mu - e$ decay chain. The result were in agreement with Wu's experiment putting another nail in the coffin of P conservation. P violation allows pseudoscalar terms in the Hamiltonian; hence Eq. (2.12) is generalised as

$$\mathscr{H}^{\beta} = \sum_i (\bar{\psi}_p \Gamma_i \psi_n)(\bar{\psi}_e (G_i + G'_i \gamma_5) \Gamma_i \psi_\nu) + h.c., \tag{2.21}$$

where 10 arbitrary constants are now involved.

2.3 TWO-COMPONENTS NEUTRINO IN WEAK INTERACTIONS

Soon after all this, in 1957, Landau [10] and Salam [11] as well as Lee and Yang [12] proposed the two-component neutrino hypothesis. This is consistent with the assumption that the neutrino is a massless particle with a definite helicity. Write the neutrino field operator in the form

$$\nu(x) = \nu_L(x) + \nu_R(x), \tag{2.22}$$

where ν_L and ν_R are the left- and right-handed components of the field $\nu(x)$, respectively. They are defined, as usual, by

$$\nu_L(x) = \frac{1-\gamma_5}{2}\nu(x), \tag{2.23}$$

$$\nu_R(x) = \frac{1+\gamma_5}{2}\nu(x). \tag{2.24}$$

The OPE of $\nu(x)$ is given by

$$\nu(x) = \int N_p \left[u^r(p)\; c_r(p)\; e^{ipx} + u^r(-p)\; d_r^\dagger(p)\; e^{-ipx}\right] d^3p, \tag{2.25}$$

where

$$N_p = \frac{1}{(2\pi)^{3/2}} \frac{1}{\sqrt{2p^0}} \tag{2.26}$$

and $c_r(p)$ and $d_r^\dagger(p)$ are the neutrino annihilation and creation operators, respectively, with momentum p and helicity r. The spinors $u^r(p)$ and $u^r(-p)$

satisfy the Dirac equations,

$$\not{p}u^r(p) = 0, \tag{2.27}$$
$$\not{p}u^r(-p) = 0, \tag{2.28}$$

where the neutrino is assumed to be massless. Here, the spinors $u^r(p)$ are eigen-functions of the helicity operator,

$$(\Sigma \cdot K)u^r(p) = ru^r(p), \tag{2.29}$$

where $r = \pm 1$ and $K = \frac{\vec{p}}{|\vec{p}|}$. Note that the matrix Σ is defined, in Dirac-Pauli representation, as

$$\Sigma = \gamma_5 \vec{\alpha} = \begin{pmatrix} \vec{\sigma} & 0 \\ 0 & \vec{\sigma} \end{pmatrix}. \tag{2.30}$$

For a massless particle, Eq. (2.29) takes the simple form:

$$\gamma_5 u^r(p) = ru^r(p). \tag{2.31}$$

Thus, in the case of a massless particle, the eigenstates of the γ_5-matrix (called chirality) are states with definite helicity. Therefore, the operator expansions of ν_L and ν_R are given by

$$\nu_{L,(R)}(x) = \int N_p[u^{-1(1)}(p)\; C_{-1(1)}(p)\; e^{-ipx} + u^{+1(-1)}(p)\; d^\dagger_{1(-1)}(p)\; e^{ipx}]d^3p. \tag{2.32}$$

The two-component neutrino hypothesis consists of the assumption that the operator $\nu_L(x)$ or $\nu_R(x)$ enters into the weak interaction Hamiltonian. As can be seen from Eq. (2.31), the neutrino in this case is a particle with negative (positive) helicity while the antineutrino is a particle with positive (negative) helicity. If the neutrino is two-component, then the constants G'_i and G_i in Eq. (2.21) are related, $G'_i = G_i$, and the Hamiltonian acquires the form

$$\mathcal{H}^\beta = 2\Sigma_i G_i(\bar{p}\Gamma^i n)(\bar{e}\Gamma^i \nu_L) + h.c., \tag{2.33}$$

with five coupling constants.

2.4 $V\!-\!A$ THEORY

The next important step in constructing the weak interaction Hamiltonian was made in 1958 by Feynman and Gell-Mann [13] as well as Marshak and Sudarshan [14]. These authors proposed that not only the neutrino field operator but also the operators of all other fermion fields enter into the weak interaction Hamiltonian through the left-handed components, $\psi_L = \frac{1}{2}(1-\gamma_5)\psi$, only. Consider, for example, the expressions $\bar{e}_L\Gamma^i\nu_L$, it is clear that

$$\bar{e}_L\Gamma^i\nu_L = \bar{e}\frac{1+\gamma_5}{2}\Gamma^i\frac{1-\gamma_5}{2}\nu. \tag{2.34}$$

Furthermore, we obtain

$$\frac{1+\gamma_5}{2}\frac{1-\gamma_5}{2} = \frac{1+\gamma_5}{2}\sigma_{\alpha\beta}\frac{1-\gamma_5}{2} = \frac{1+\gamma_5}{2}\gamma_5\frac{1-\gamma_5}{2} = 0. \tag{2.35}$$

Thus, if only the left-handed components of fermion fields are involved in the Hamiltonian, then the scalar, tensor and pseudoscalar interactions are forbidden. The components of the vector and axial-vector do not vanish and will survive with the following relation:

$$\frac{1+\gamma_5}{2}\gamma_\alpha\gamma_5\frac{1-\gamma_5}{2} = \frac{1+\gamma_5}{2}\gamma_\alpha\frac{1-\gamma_5}{2} = -\gamma_\alpha\frac{1-\gamma_5}{2}. \tag{2.36}$$

In this regard, the Hamiltonian must be a combination of the $V-A$ couplings, which can be written as

$$\begin{aligned}\mathcal{H}^\beta &= \frac{G_F}{\sqrt{2}}4(\bar{\psi}_{p_L}\gamma_\mu\psi_{n_L})(\bar{\psi}_{e_L}\gamma^\mu\psi_{\nu_L}) + h.c.\\ &= \frac{G_F}{\sqrt{2}}(\bar{\psi}_p\gamma_\mu(1-\gamma_5)\psi_n)(\bar{\psi}_e\gamma^\mu(1-\gamma_5)\psi_\nu) + h.c.\end{aligned} \tag{2.37}$$

The Hamiltonian in Eq. (2.37) does not only successfully describe β-decay of a nucleon but also describes any other process involving electrons, neutrinos and hadrons in which the lepton number is conserved:

$$\bar{\nu} + p \to e^+ + n, \tag{2.38}$$

$$\pi^+ \to e^+ + \nu. \tag{2.39}$$

It is also noticeable that $\bar{\nu} + n \to e^- + p$ is not included in this Hamiltonian.

2.5 CONSERVATION OF LEPTON NUMBER

We have assumed that the operator $\nu(x)$ is a Dirac operator, *i.e.*, $\nu^c = C\bar{\nu}^T \neq \nu$, so that the neutrino and antineutrino are different particles. The charge, by which the neutrino is distinguished from the antineutrino, is called the lepton number. The lepton numbers of the neutrino and electron are equal to 1, while the lepton numbers of the antineutrino and positron are equal to -1. The lepton numbers of all other particles are equal to zero. The Hamiltonian in Eq. (2.37) has a global gauge invariance under the following transformations:

$$\begin{aligned} \nu &\rightarrow e^{i\alpha}\nu, \qquad e \rightarrow e^{i\alpha}e, \\ p &\rightarrow p, \qquad n \rightarrow n. \end{aligned} \tag{2.40}$$

So far, only weak processes have been discussed, in which an electron, neutrino and hadrons are involved. We shall now consider weak processes in which a muon, neutrino and hadrons take part, *i.e.*,

$$\mu^- + p \rightarrow \nu_\mu + n. \tag{2.41}$$

In accordance to the above invariance, the total lepton charge is conserved, *i.e.*, $\Sigma_i L_i =$ constant. The law of lepton charge conservation is confirmed by available experimental data. For instance, the following muon decay processes are forbidden: $\mu^+ \rightarrow e^+\gamma$ and $\mu^+ \rightarrow e^+ + e^+ + e^-$. The ratios of the probabilities of these processes to the probability of the allowed decay $\mu^+ \rightarrow e^+ + \nu_e + \bar{\nu}_\mu$ are about $R < 4.9 \times 10^{-11}$ and $R < 1 \times 10^{-12}$, respectively.

2.5.1 Lepton Universality

The simplest extension to the Fermi coupling was proposed by Feynman and Gell-Mann in [13], as

$$\mathscr{H} = \frac{4G_F}{\sqrt{2}}(\bar{p}_L\gamma^\alpha n_L)[\bar{e}_L\gamma_\alpha\nu_L + \bar{\mu}_L\gamma_\alpha\nu_L] + h.c. \tag{2.42}$$

Also, they introduced the weak-hadron currents,

$$j^\alpha = 2[\bar{p}_L\gamma^\alpha n_L + \bar{\nu}_{eL}\gamma^\alpha e_L + \bar{\nu}_{\mu L}\gamma^\alpha \mu_L], \tag{2.43}$$

$$j^+_\alpha = 2[\bar{n}_L\gamma_\alpha p_L + \bar{e}_L\gamma_\alpha\nu_{eL} + \bar{\mu}_L\gamma_\alpha\nu_{\mu L}]. \tag{2.44}$$

Here $\mu - e$ universality was assumed, *i.e.*, both μ and e terms have the same coupling; hence, the general expression of the Hamiltonian in terms of j^α and j_α^+ is given by

$$\mathscr{H} = \frac{G_F}{\sqrt{2}}\, j^\alpha j_\alpha^+. \tag{2.45}$$

It is clear that this Hamiltonian (current-current interaction) includes the β-decay and μ-decay as the non-diagonal terms. There are also the additional diagonal terms:

$$\mathscr{H}^d = \frac{G_F}{\sqrt{2}}[(\bar{p}_L\gamma^\alpha n_L)(\bar{n}_L\gamma_\alpha p_L)+(\bar{\nu}_{eL}\gamma^\alpha e_L)(\bar{e}_L\gamma_\alpha\nu_{eL})+(\bar{\nu}_{\mu L}\gamma^\alpha\mu_L)(\bar{\mu}_L\gamma_\alpha\nu_{\mu L})]. \tag{2.46}$$

It is an important check for this current-current theory to investigate the processes described by these diagonal terms. The first term is responsible for a pure lepton process like the following scattering

$$\bar{\nu}_e + e \to e + \bar{\nu}_e. \tag{2.47}$$

There exists a single experiment with antineutrinos from a reactor in which this process has been observed (by Reines and others [15]). At this point, it is not obvious whether a distinct neutrino field ν_ν has to be added to the theory but let us take it, with a pinch of salt, as the more general case.

2.5.2 Neutrino Flavours

In 1962, an experimental evidence for the double nature of neutrino was found [15], using neutrino beam methods as follows. Prepare a beam of π^+ mesons from an accelerator, for which we predict that it decays as $\pi^+ \to \mu^+ + \nu_\mu$ (this is 10^4 of the probability of the decay $\pi^+ \to e^+ + \nu_e$). If, along the path of the beam at a distance greater than the decay length of the pions there is a shield wide enough to absorb all the decay muons, then after the shield we will have a ν_μ beam. Let us assume that a neutrino detector (a device which registers particles created in neutrino processes) is placed along the ν_μ beam, so that, if ν_μ and ν_e are different particles, then within the detector it will be possible to observe muons produced in the process

$$\nu_\mu + N \;\to\; \mu^- + \text{hadrons}. \tag{2.48}$$

If ν_μ and ν_e are identical particles, then, within the detector, both muons and electrons would be produced, thus also the process

$$\nu_\mu + N \;\to\; \bar{e} + \text{hadrons} \tag{2.49}$$

will be registered. Further, if so, the number of muons must be equal to the number of electrons. This experiment showed that ν_μ and ν_e are different particles.

In accordance with the results of this and other experiments we introduce two lepton numbers, *i.e.*, the electron number L_e and the muon number L_μ. Assume that the electron and electron neutrino (the neutrino which is emitted together with a positron) both have L_e values equal to 1 and L_μ values equal to zero. Assume also the muon numbers of the μ^- state and of the muon neutrino (the neutrino which is emitted together with a μ^+) to be equal to unity. Finally set the lepton numbers of hadrons (h) and photons (γ) to be equal to zero. The lepton numbers of these particles are summarised in Tab. 2.1.

	ν_e , e^-	ν_μ , μ^-	h , γ
L_e	1	0	0
L_μ	0	1	0

Table 2.1 Lepton number assignments.

It is now clear that Eq. (2.42) is engineered to obey lepton number conservation and lepton universality, *i.e.*, $\sum_i L_e^{(i)}$ = constant and $\sum_i L_\mu^{(i)}$ = constant.

2.6 WEAK INTERACTION OF STRANGE PARTICLES

Until now, we only discussed lepton-hadron processes in which the strangeness of the hadron remains unchanged. Strange particles were discovered long before the current-current theory. It was noticed that strangeness is conserved during strong and EM interactions, however, not during weak interactions. Feynman and Gell-Mann introduced into the charged current a term responsible for the decay of these strange particles. Consider decays of strange particles in which lepton neutrino pairs are produced, *i.e.*, $h_i \to h_f + \ell + \nu_\ell$, where h_i is the initial hadron, h_f is the final hadron and $\ell = e, \mu$. The following processes are examples of this type of decays:

$$K^+ \rightarrow e^+ + \nu_e, \tag{2.50}$$

$$K^+ \rightarrow \pi^0 + e^+ + \nu_e, \tag{2.51}$$

$$\Lambda \rightarrow p + e^- + \bar{\nu}_e, \tag{2.52}$$

$$\Sigma \rightarrow n + e^- + \bar{\nu}_e, \tag{2.53}$$

$$\Xi \rightarrow \Lambda + e^- + \bar{\nu}_e. \tag{2.54}$$

As a result of the study of these decays, the following empirical rules were formulated.

1. Decays with $\mid \Delta S \mid \geq 2$ are forbidden, *i.e.*, $|\Delta S| \leq 1$.
2. The relation $\Delta Q = \Delta S$ holds, where ΔQ and ΔS are the change of electric charge and strangeness between the initial and final hadrons.
3. The effective coupling constant for processes changing strangeness is about four times as small as that of processes that conserve strangeness.

We shall now present some experimental data to illustrate these rules.

2.6.1 Experimental Verification

From rule no. 1, the following process should be forbidden since $\Delta S = 2$:

$$\Xi^- \rightarrow p + \pi^- + e^- + \bar{\nu}_e\,. \tag{2.55}$$

Experimentally, the ratio of this probability to the total decay probability of the Ξ^- hyperon is $R \leq 9 \times 10^{-4}$.

From rule no. 2, the following decay with $\Delta Q = -1$ and $\Delta S = 1$ should be forbidden:

$$\Sigma^+ \rightarrow n + e^+ + \nu_e\,. \tag{2.56}$$

Experimentally, the ratio of this probability to the total decay probability of the Σ^+ hyperon is $R \leq 0.5 \times 10^{-5}$. Furthermore, the ratio of this decay can be compared to the following allowed decay $\Sigma^- \rightarrow n + e^- + \bar{\nu}_e$, as

$$\frac{\Gamma_{\Sigma^+ \rightarrow n e^+ \nu_e}}{\Gamma_{\Sigma^- \rightarrow n e^- \bar{\nu}_e}} \leq 0.04. \tag{2.57}$$

By the same rule, the $K^+ \to \pi^+ + \pi^+ + e^- + \bar{\nu}_e$ decay is forbidden while the $K^+ \to \pi^+ + \pi^- + e^+ + \nu_e$ decay is allowed. As expected, the allowed decay has a much larger probability:

$$\frac{\Gamma_{K^+\to\pi^+\pi^+ e\bar{\nu}_e}}{\Gamma_{K^+\to\pi^+\pi^- e^+\nu_e}} \leq 0.01. \tag{2.58}$$

To illustrate the suppression of decays with $\mid \Delta S \mid = 1$ compared to decays with $\Delta S = 0$, we consider as an example the following decay processes: $K^+ \to \mu^+ + \nu_\mu$ and $\pi^+ \to \mu^+\nu_\mu$. From experimental data the ratio of probabilities was obtained as

$$\frac{\Gamma_{K^+}}{\Gamma_{\pi^+}} \simeq 0.27. \tag{2.59}$$

2.7 WEAK INTERACTION OF QUARKS

In 1963, Cabibbo found an expression for the charged hadron current which allowed the description of all data on strange particle decays [16]. The Cabibbo current is based on the $SU(3)$ symmetry and satisfies strictly the rule $\Delta S = 0$ and $\mid S \mid \leq 1$. We shall construct this current of quark fields.

The quark hypothesis (of a subatomic particle which fractional electric charge, which field transforms under the fundamental $SU(3)_C$ group representation of colour, see later on) was proposed independently by Gell-Mann and Zweig in 1964 to explain particle spectroscopy. The flavour properties of quarks are given in Tab. 2.2, where Q is the electric charge, I_3 is the weak isospin projection, B is the baryon number and the strangeness S is related to the above quantum numbers by the Gell-Mann-Nishijima relation:

$$Q = I_3 + \frac{1}{2}(B + S)\,. \tag{2.60}$$

	I_3	B	S	Q
u	$\frac{1}{2}$	$\frac{1}{3}$	0	$\frac{2}{3}$
d	$-\frac{1}{2}$	$\frac{1}{3}$	0	$-\frac{1}{3}$
s	0	$\frac{1}{3}$	-1	$-\frac{1}{3}$

Table 2.2 Flavour properties of the up, down and strange quarks.

We shall assume that only the left-handed components of the quark fields $q_L(x)$ enter the charged current $q_L(x) \equiv \frac{1-\gamma_5}{2} q(x)$. The simplest current to be constructed for the quark field is the following one:

$$\bar{q'}_L(x)\gamma_\alpha q_L(x) = \bar{q}'\gamma_\alpha \frac{1-\gamma_5}{2} q(x). \tag{2.61}$$

Let us construct the charged current that does not change strangeness. Clearly, such a current must involve the fields of the u and d quarks

$$J_\alpha(\Delta S = 0) = \bar{u}_L \gamma_\alpha d_L. \tag{2.62}$$

To find the strangeness changing current, we note that it must obey $\Delta S = \Delta Q$. For a current in the form $\bar{q'}_L(x)\gamma_\alpha q_L(x)$, it is obvious that this current is uniquely:

$$J_\alpha(\Delta S = 1) = \bar{u}_L \gamma_\alpha s_L\,. \tag{2.63}$$

The total hadron current may then be written in the form

$$(J_\alpha)_c = a J_\alpha(\Delta S = 0) + b J_\alpha(\Delta S = 1)\,. \tag{2.64}$$

From experiment, we know that the coefficient b is much smaller than the coefficient a. Cabibbo proposed that $a^2 + b^2 = 1$, so that the expression can be parametrised by one angle θ_c, known as Cabibbo angle (such that $a = \cos\theta_c$ and $b = \sin\theta_c$). In this case, the current $(J_\alpha)_c$ is given by

$$(J_\alpha)_c = \bar{u}_L \gamma_\alpha d', \tag{2.65}$$

where

$$d' \equiv d_L \cos\theta_c + s_L \sin\theta_c\,. \tag{2.66}$$

The weak charged current can now be written as

$$J_\alpha = 2[\bar{\nu}_{eL}\gamma_\alpha e_L + \bar{\nu_\mu}_L \gamma_\alpha \mu_L + \bar{u}_L \gamma_\alpha d']\,. \tag{2.67}$$

2.7.1 A New Quark

The expression in Eq. (2.67) has a familiar $V - A$ structure (only the left-handed field components are present). However there is a clear discrepancy between lepton and hadron terms. This is actually a result of constructing the theory with four leptons but three quarks. To tackle this asymmetry, a fourth quark (charm, c) was proposed in 1964 [17], with properties given in Tab. 2.3.

	I_3	B	S	Q
c	0	$\frac{1}{3}$	0	$\frac{2}{3}$

Table 2.3 Flavour properties of the charm quark.

In the Cabibbo current, a certain combination between d and s quarks is present in d' in Eq. (2.66). It is natural to assume the additional current term orthogonal to the above combination as follows:

$$s'_L = -d_L \sin\theta_c + s_L \cos\theta_c \,. \tag{2.68}$$

Under the usual assumption that only left-handed field components enter the Hamiltonian, we reach the following expression for the additional term in the charged current:

$$(J_\alpha)_{\text{GIM}} = \bar{c}_L \gamma_\alpha s'_L \,. \tag{2.69}$$

This term was added by Sheldon Lee Glashow, John Iliopoulos and Luciano Maiani, which is known as GIM mechanism [18].

The existence of a new quark predicted the existence of a new set of composite particles that include the c-quark in their formation. The first of these particles was discovered in 1974[1] [19,20], confirming the existence of the charm quark. The charged current is now modified to be

$$J_\alpha = \bar{\nu}_{eL} \gamma_\alpha e_L + \bar{\nu}_{\mu L} \gamma_\alpha \mu_L + \bar{u}_L \gamma_\alpha d' + \bar{c}_L \gamma_\alpha s', \tag{2.70}$$

where d' and s' are those in Eqs. (2.66) and (2.68), respectively. A slight discrepancy between the lepton and hadron terms still remains. Notice how the d and s quarks appear in a mixed form unlike the neutrinos in the lepton terms. This can be remedied by considering ν_e and ν_μ as orthogonal combination of

[1] It was the J/ψ meson, which is a bound state of $c\bar{c}$.

some other fields:

$$\nu_e(x) = \nu_1(x)\cos\theta' + \nu_2(x)\sin\theta', \tag{2.71}$$

$$\nu_\mu(x) = -\nu_1(x)\sin\theta' + \nu_2(x)\cos\theta', \tag{2.72}$$

with ν_1 and ν_2 being the field operators of the neutrinos with masses m_1 and m_2, respectively. Here, θ' is a parameter analogous to the Cabibbo angle. The neutrino fields are now written as

$$\begin{pmatrix} \nu_e \\ \nu_\mu \end{pmatrix} = \begin{pmatrix} \cos\theta' & \sin\theta' \\ -\sin\theta' & \cos\theta' \end{pmatrix} \begin{pmatrix} \nu_1 \\ \nu_2 \end{pmatrix}. \tag{2.73}$$

2.7.2 A New Lepton and Another New Quark

More recent experimental data indicated that Eq. (2.70) does not represent the complete expression for the charged current since the τ lepton has been discovered with a mass of 1784 MeV [21]. This data can be described by adding a term that treats τ on the same footing as e and μ,

$$(J_\alpha)_\tau = \bar{\nu_\tau}_L \gamma_\alpha \tau_L, \tag{2.74}$$

where ν_τ is the field operator of the neutrino with the new lepton charge.

If a lepton-hadron symmetry exists, one can expect another hadron term to show up in the charged current. This additional piece can reasonably be constructed in a similar form to Eqs. (2.65) and (2.69),

$$(J_\alpha)' = \bar{t}_L \gamma_\alpha b_L, \tag{2.75}$$

where t and b are the quark fields of the new particles (called bottom and top quarks) that have the properties listed in Tab. 2.4.

	I_3	B	S	Q
b	0	$\frac{1}{3}$	0	$-\frac{1}{3}$
t	0	$\frac{1}{3}$	0	$\frac{2}{3}$

Table 2.4 Properties of the third generation of quarks.

The first experimental indication of the existence of a fifth quark was obtained in 1977 by Fermilab. By colliding e^+e^- beams, a set of new particles, named Y (or Υ) particles, was discovered. These particles are very heavy (with $m_Y \simeq 9.46$ GeV and $m_{Y'} \simeq 10.01$ GeV) [22]. The available data are consistent with Y particles being bound states of a $b\bar{b}$ system. In addition, the top quark was discovered in 1995 by the CDF [23] and D0 [24] experiments at FNAL (see Chapter 13). It has an electric charge of $2/3\ e$ and a mass of 172.76 ± 0.3 GeV. We thus conclude the section by writing the final form for the charged weak current:

$$J_\alpha = \bar{\nu_e}_L \gamma_\alpha e_L + \bar{\nu_\mu}_L \gamma_\alpha \mu_L + \bar{\tau \mu}_L \gamma_\alpha \tau_L + \bar{u}_L \gamma_\alpha d' + \bar{c}_L \gamma_\alpha S' + \bar{t}_L \gamma_\alpha b_L. \quad (2.76)$$

2.8 INTERMEDIATE GAUGE BOSON

The point-like interaction described by the $V - A$ theory proposed by Fermi had some serious problems in predictions at high energies. The assumption of an intermediate boson $W^\pm$, carrier of the weak force, could alleviate these problems. A fortiori, it is the case that the weak interaction Hamiltonian in Eq. (2.45) was constructed in an analogous way to the EM theory. Thus, the existence of an intermediate gauge boson will help put the two theories on equal footing since QED has photons as virtual particles that mediate the interaction between the currents ($\mathscr{L}_I^{\text{EM}} \sim e\ \bar{e}\gamma_\alpha e A^\alpha$). The need for this particle in the weak interaction theory becomes evident if we try to formulate it as a gauge theory, this will be discussed in a subsequent chapter.

If such a particle exists, it must be a vector particle (from Lorentz invariance). We also notice that it must be charged in order to explain weak processes as those given in Fig. 2.1. To solve the short-range problem of the weak interaction, it must be massive as well. (The mechanism by which it gains its mass will be discussed later.)

If such a vector boson exists, the fundamental weak interaction Lagrangian should have the form

$$\mathscr{L} = -\frac{g}{2\sqrt{2}} J^\alpha W_\alpha + h.c., \quad (2.77)$$

where J^α is the charged weak current and W_α the field operator of the $W^\pm$ boson. At low energies, $E \ll M_{W^\pm}$, the mass of the vector boson can be estimated as $M_{W^\pm}^2 \simeq \frac{\sqrt{2}g^2}{8}$. Further, if the $W^\pm$ boson exists, there is hope in constructing a unified theory of weak and EM interactions. That was actually

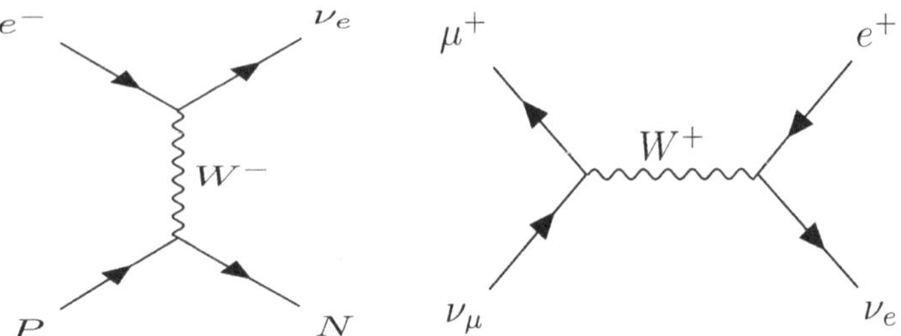

Figure 2.1 Weak processes mediated by a $W^\pm$ boson.

the case, the EW theory was indeed constructed by Weinberg and Salam in 1967 (see later on).

Only through this unification, a renormalisable theory of the weak interactions was obtained. Before which, the situation was a quagmire. Although the current-by-current interaction and intermediate gauge boson perfectly described the enormous bulk of data, higher order effects in perturbation theory had no sense, in particular, it was impossible to show that their contribution was small.

2.8.1 Unitarity Violation

The standard calculation of the cross section in the $V - A$ theory for the simple scattering process $e\nu_e \to \nu_e$ can be shown to be [25]

$$\sigma = \frac{G_F^2}{3\pi}(s - m^2)\left[1 - \left(\frac{m^2}{s}\right)^3\right], \tag{2.78}$$

where s is the standard Mandelstam variable. So that, at high energies $E \gg m$, one finds $\sigma \approx G_F^2 s$. This can be motivated on a dimensional analysis ground. As pointed out before, the Fermi constant has dimensions $[G_F] = M^{-2}$, the cross section has dimensions $[\sigma] = M^{-2}$ and it should be proportional to the square of the matrix element $\sigma \sim G_F^2$, so the cross section must depend on some variable that have dimension M^2 hence the variable $s \equiv E^2$ appears.

This divergent behaviour at high energies has serious implications on the unitarity of the theory. Unitarity is the requirement that the probability of all final states of a given initial state must add up to one. It is obvious that, for sufficiently large s, the probability of this interaction will exceed 1, violating this principle. This will be the basis of our discussion of Chapter 12.

CHAPTER 3

Non-Abelian Gauge Theories

In this chapter, we will discuss the idea of non-Abelian gauge theories, which involve transformations under generators that do not commute with each other so that the latter will depend on the order of the group. First, we will discuss the most general construction to understand how fields transform under such transformations and their properties. Second, we study two special cases, namely, $SU(2)$ and $SU(3)$, which are pertinent to the construction of the SM. Following the same strategy of Abelian gauge theories, starting from a Lagrangian that is invariant under non-Abelian transformations, we extend these to be local and construct a gauge invariant Lagrangian under such conditions.

3.1 CONSTRUCTION OF NON-ABELIAN GAUGE THEORIES

A Lie group is a representation group that has a finite number of parameters, this means that for every element i of a group G we have a matrix U_i that satisfies the same multiplication rules as the corresponding elements of G. The number of independent parameters defines the dimensionality of the group.

The non-Abelian groups $SO(N)$ and $SU(N)$ are of particular importance in particle physics. In fact, $SO(N)$ is the group of real rotations in N dimensions ($N > 2$). If A is matrix in $SO(N)$ then it must satisfy $A^T A = AA^T = 1$ and has a unit determinant: these conditions restrict the number of independent parameters of A to $\frac{1}{2}N(N-1)$. Instead, $SU(N)$ is a group of $N \times N$ unitary matrices ($N > 2$) of unit determinant, *i.e.*, if A is matrix in $SU(N)$

DOI: 10.1201/9780429443015-3

then it must satisfy $A^\dagger A = AA^\dagger = 1$ and $\det(A) = 1$: these conditions restrict the independent parameters of A to $N^2 - 1$.

Let us now consider field transformations under such groups,

$$\psi(x) \to \psi'(x) = U\psi(x), \tag{3.1}$$

where ψ is an array of different fields written as column vector $\psi = (\psi_1, \psi_2, \psi_3, \ldots)^T$. We may write the action of the group on every element of it as

$$\psi(x) \to \psi_i'(x) = U_{ij}\psi_j(x). \tag{3.2}$$

The matrices U can be written in exponential form

$$U = \exp(-ig\theta^\alpha t_\alpha), \tag{3.3}$$

where g is a constant (which will represent a coupling constant in particle physics interactions) and the matrices t_α's are called the generators of the group while the corresponding θ^α's constitute a set of linearly independent real parameters. Thus, the number of generators equals the number of independent parameters. We note that, from the definition of $SO(N)$ and $SU(N)$, we can prove the following: the generators of $SO(N)$ must consist of $N \times N$ real and anti-symmetric matrices and the generators of $SU(N)$ must be anti-Hermitian $N \times N$ matrices, which, furthermore, must be traceless in order to have a unit determinant.

To get a mathematical framework for such groups, consider infinitesimal transformations where the parameters θ^α are small, so that

$$\begin{aligned} U &= 1 - ig\,\theta^\alpha\, t_\alpha + O(\theta^2), \\ U^{-1} &= 1 + ig\,\theta^\alpha\, t_\alpha + O(\theta^2). \end{aligned} \tag{3.4}$$

Because the matrices U defined in Eq. (3.2) form a group, products of these matrices must be expressible into the same exponential form (closure property). This leads to an important relation between the generators t_α implying that they must close under commutation,

$$[t_a, t_b] = if^c_{ab}t_c, \tag{3.5}$$

where f^c_{ab} are called structure constants.

3.1.1 Local non-Abelian Gauge Group

Consider the extension of the group G to a group of local gauge transformations. This means that the parameters of G will become functions of the space-time coordinates x^μ,

$$\psi(x) \to \psi'(x) = U(x)\psi(x). \tag{3.6}$$

Thus, the derivative of a field leads to

$$\partial_\mu \psi \to (\partial_\mu \psi)' = U(x)\partial_\mu \psi + (\partial_\mu U)\psi. \tag{3.7}$$

We note that the presence of the second term on the right-hand side of $\partial_\mu \psi$ spoils the invariance under the local group: *i.e.*, derivatives of fields do not transform covariantly. The general philosophy of gauge theory is imposing local invariance, this can be accommodated by replacing $\partial_\mu \psi$ by a so-called covariant derivative D_μ which constitutes a covariant quantity when applied to ψ:

$$D_\mu \psi \to (D_\mu \psi)' = U(x) D_\mu \psi. \tag{3.8}$$

We will construct this covariant derivative as an Abelian transformation by introducing an auxiliary "gauge field" W_μ,

$$D_\mu \psi = \partial_\mu \psi - ig W_\mu \psi, \tag{3.9}$$

where W_μ is a matrix of the type generated by an infinitesimal gauge transformation, *i.e.*, W_μ can be decomposed into a linear combination of the generators t_α:

$$W_\mu = \sum W_\mu^\alpha t_\alpha = \vec{W}_\mu \cdot \vec{t}. \tag{3.10}$$

Let us look for the desired transformation on W_μ that satisfies the condition in Eq. (3.8). By using Eq. (3.9), one can show the following

$$\begin{aligned} W_\mu \psi' &= \frac{1}{g}(\partial_\mu \psi' - (D_\mu \psi)') \\ &= \frac{1}{g}(\partial_\mu (U\psi) - U D_\mu \psi \end{aligned}$$

$$= \frac{1}{g}(U\partial_\mu\psi + (\partial_\mu U)\psi - UD_\mu\psi)$$
$$= \left(UW_\mu U^{-1} - i\frac{1}{g}(\partial_\mu U)U^{-1}\right)\psi'. \tag{3.11}$$

This implies that

$$W_\mu \to W'_\mu = UW_\mu U^{-1} - i\frac{1}{g}(\partial_\mu U)U^{-1}. \tag{3.12}$$

Now, for infinitesimal translations as in Eqs. (3.4) and (3.10), Eq. (3.12) can be written as

$$W^a_\mu \to (W^a_\mu)' = W^a_\mu + f^a_{bc}\theta^b W^c_\mu - \frac{1}{g}\partial_\mu\theta^a. \tag{3.13}$$

This result differs from the transformation law of Abelian gauge fields by the presence of the term $f^a_{bc}\theta^b W^c_\mu$. In this regard, the W^a's transform as a field in the adjoint representation and their number is determined by the dimension of the Lie algebra associated with the gauge group. It is also important to note that ordinary derivatives commute whereas two covariant derivatives do not commute with each other:

$$[D_\mu, D_\nu]\psi = -ig(\partial_\mu W_\nu - \partial_\nu W_\mu + g[W_\mu, W_\nu])\psi. \tag{3.14}$$

Therefore, the anti-symmetric stress tensor $G_{\mu\nu}$ can be defined as

$$G_{\mu\nu} \equiv \frac{i}{g}[D_\mu, D_\nu] = \partial_\mu W_\nu - \partial_\nu W_\mu + g[W_\mu, W_\nu]. \tag{3.15}$$

Using the transformation properties for D_μ, we can prove the following relation:

$$([D_\mu, D_\nu]\psi)' = U([D_\mu, D_\nu]\psi). \tag{3.16}$$

From Eqs. (3.15) and (3.16), one can find the gauge transformation of the field strength $G_{\mu\nu}$ as follows:

$$G_{\mu\nu} \to (G_{\mu\nu})' = UG_{\mu\nu}U^{-1}. \tag{3.17}$$

Since W_μ is Lie-algebra valued, the field strength in Eq. (3.15) is also Lie algebra valued, *i.e.*, $G_{\mu\nu}$ can also be decomposed in terms of the group generators

t_a,

$$G_{\mu\nu} = \sum G^a_{\mu\nu} t_a = \vec{G}_{\mu\nu} \cdot \vec{t}, \tag{3.18}$$

where $G^\alpha_{\mu\nu}$ is given by

$$G^a_{\mu\nu} = \partial_\mu W^a_\nu - \partial_\nu W^a_\mu + g f^a_{bc} W^b_\mu W^c_\nu. \tag{3.19}$$

Unlike the Abelian field strength, $G_{\mu\nu}$ is not invariant under the gauge transformation

$$G_{\mu\nu} \to (G_{\mu\nu})' = G_{\mu\nu} + [\theta^a t_a, G_{\mu\nu}]. \tag{3.20}$$

However, the term $G^a_{\mu\nu} G^{a\mu\nu}$ will be invariant and will represent a 'kinetic term' of the theory, as we will see later. In summary, the outline of a non-Abelian gauge theory is looking for local field transformations that preserve the global invariance, this led us to introduce the gauge field W_μ and the strength tensor for this field representing a kinetic term. In the next section, we examine two examples of non-Abelian groups of particular interest, $SU(2)$ and $SU(3)$.

3.2 GAUGE THEORY OF $SU(2)$

Let us first consider the Lagrangian for a free particle with mass m

$$\mathscr{L}_0(\psi, \bar{\psi}) = \bar{\psi}(x)(i\gamma^\mu \partial_\mu - m)\psi(x), \tag{3.21}$$

where $\psi(x)$ is a set of N spinor fields and the γ^μ's are the Dirac matrices. The fields ψ and $\bar{\psi}$ transform under U which belong to a certain group G as follows:

$$\psi(x) \to \psi'(x) = U\psi(x), \qquad \bar{\psi}(x) \to \bar{\psi}(x)' = \bar{\psi}(x)U^\dagger. \tag{3.22}$$

This group consists of all 2×2 unitary matrices with unit determinant so it has 3 generators $((2)^2 - 1)$ called t_a, $a = 1, 2, 3$. (The first construction of a non-Abelian gauge theory with an $SU(2)$ group was presented by Yang and Mills [26].) Such matrices can be written in the usual exponential form

$$U(\theta) = \exp(-ig\theta^a t_a) \quad (a = 1, 2, 3), \tag{3.23}$$

where g is coupling and t_a are the three generators of $SU(2)$ which can be represented by the isotopic spin matrices

$$t_a = \frac{1}{2}\tau_a, \tag{3.24}$$

with τ_a being the isospin (effectively, Pauli) matrices

$$\tau_1 = \begin{pmatrix} 0 & 1 \\ 1 & 0 \end{pmatrix}, \quad \tau_2 = \begin{pmatrix} 0 & -i \\ i & 0 \end{pmatrix}, \quad \tau_3 = \begin{pmatrix} 1 & 0 \\ 0 & -1 \end{pmatrix}. \tag{3.25}$$

Consequently the generators t_a satisfy the commutation relations

$$[t_a, t_b] = i\epsilon_{abc}t_c \Rightarrow \left[\frac{\tau_a}{2}, \frac{\tau_b}{2}\right] = i\epsilon_{abc}\frac{\tau_c}{2}, \tag{3.26}$$

where ϵ_{abc} is the Levi-Civita symbol.

The fundamental representation of $SU(2)$ is a doublet, $\psi = \begin{pmatrix} \psi_1 \\ \psi_2 \end{pmatrix}$. The Lagrangian in Eq. (3.21) is invariant under a global $SU(2)$ transformation, but for a local (or gauge) symmetry one must use the technique developed in the previous section by replacing the ordinary derivative with a covariant derivative D_μ,

$$\mathscr{L}_0(\psi, \bar{\psi}) = \bar{\psi}(x)(i\gamma^\mu D_\mu - m)\psi(x), \tag{3.27}$$

which is given by $D_\mu \equiv \partial_\mu - igW_\mu$, wherein we have introduced a gauge field W_μ that belongs to the space of the transformation group, $W^\mu = \vec{W}_\mu \cdot \vec{t} = \sum W_\mu^a t_a$. Thus, the Lagrangian is given by

$$\mathscr{L}(\psi, \bar{\psi}) = \bar{\psi}(x)(i\gamma^\mu \partial_\mu - m)\psi(x) - \frac{1}{2}igW_\mu^a \bar{\psi}\gamma^\mu \tau_a \psi. \tag{3.28}$$

The interaction term between gauge and fermions fields is given by

$$\mathscr{L}_I = \frac{1}{2}igW_\mu^a \bar{\psi}\gamma^\mu \tau_a \psi. \tag{3.29}$$

Finally, we add the kinetic term in a similar fashion to that of QED. This follows straightforwardly from the $SU(2)$ structure:

$$\vec{G}_{\mu\nu} = \partial_\mu \vec{W}_\nu - \partial_\nu \vec{W}_\mu + g[\vec{W}_\mu, \vec{W}_\nu], \tag{3.30}$$

$$G_{\mu\nu}^a = \partial_\mu W_\nu^a - \partial_\nu W_\mu^a + g\epsilon_{abc}W_\mu^b W_\nu^c, \tag{3.31}$$

$$\mathscr{L} \equiv -\frac{1}{2}\text{Tr}(G_{\mu\nu}G^{\mu\nu}) = -\frac{1}{4}G^a_{\mu\nu}G^{a\mu\nu}. \tag{3.32}$$

Therefore, the total Lagrangian of $SU(2)$ is

$$\mathscr{L}_0(\psi,\bar{\psi}) = \bar{\psi}(x)(i\gamma^\mu\partial_\mu - m)\psi(x) + \frac{1}{2}igW^a_\mu\bar{\psi}\gamma^\mu t_a\psi - \frac{1}{4}G^a_{\mu\nu}G^{a\mu\nu}. \tag{3.33}$$

It is noticeable that, similar to the Abelian case, here too the mass term of the gauge fields are forbidden by the gauge symmetry $SU(2)$.

3.3 GAUGE THEORY OF $SU(3)$

In this section we discuss $SU(3)$. This group consists of all 3×3 unitary matrices with unit determinant, it has $(3)^2 - 1 = 8$ generators which are labelled as ρ^i, $i = 1, ..., 8$. Hence, the unitary gauge matrix for $SU(3)$ can be written in exponential form as

$$U(\theta) = \exp(-ig_s\theta^i\rho_i) \quad (i = 1, ..., 8), \tag{3.34}$$

where g_s is the coupling constant of $SU(3)$ and θ^i are constant parameters. The generators ρ_i satisfy the commutation relations

$$[\rho_a, \rho_b] = i\epsilon_{abc}\rho_c. \tag{3.35}$$

The eight generators of $SU(3)$ can be represented by the Gell-Mann matrices $\rho_i = \frac{1}{2}\lambda_i$, which implies that

$$\left[\frac{\lambda_a}{2}, \frac{\lambda_b}{2}\right] = i\epsilon_{abc}\frac{\lambda_c}{2}. \tag{3.36}$$

The matrix representation of λ_a, $a = 1, ..., 8$ is given by

$$\lambda_1 = \begin{pmatrix} 0 & 1 & 0 \\ 1 & 0 & 0 \\ 0 & 0 & 0 \end{pmatrix}, \quad \lambda_2 = \begin{pmatrix} 0 & 1 & 0 \\ 1 & 0 & 0 \\ 0 & 0 & 0 \end{pmatrix}, \quad \lambda_3 = \begin{pmatrix} 0 & 1 & 0 \\ 1 & 0 & 0 \\ 0 & 0 & 0 \end{pmatrix},$$

$$\lambda_4 = \begin{pmatrix} 0 & 1 & 0 \\ 1 & 0 & 0 \\ 0 & 0 & 0 \end{pmatrix}, \quad \lambda_5 = \begin{pmatrix} 0 & 1 & 0 \\ 1 & 0 & 0 \\ 0 & 0 & 0 \end{pmatrix}, \quad \lambda_6 = \begin{pmatrix} 0 & 1 & 0 \\ 1 & 0 & 0 \\ 0 & 0 & 0 \end{pmatrix},$$

$$\lambda_7 = \begin{pmatrix} 0 & 1 & 0 \\ 1 & 0 & 0 \\ 0 & 0 & 0 \end{pmatrix}, \quad \lambda_8 = \begin{pmatrix} 0 & 1 & 0 \\ 1 & 0 & 0 \\ 0 & 0 & 0 \end{pmatrix}. \tag{3.37}$$

These generators satisfy the following useful relations:

$$\mathrm{Tr}[\lambda_a] = 0, \tag{3.38}$$

$$\mathrm{Tr}[\lambda_a \lambda_b] = 2\delta_{ab}, \tag{3.39}$$

$$\lambda^a_{ij}\lambda^a_{k\ell} = -\frac{2}{3}\delta_{ij}\delta_{k\ell} + 2\delta_{i\ell}\delta_{jk}, \qquad i,j,k,\ell = 1,2,3. \tag{3.40}$$

Applying the standard construction procedure of a gauge theory, we start with a Lagrangian invariant under a globally $SU(3)$ symmetry. To make it invariant under a local $SU(3)$ one, we replace the ordinary derivative with the covariant one

$$\mathscr{L}_0(\psi,\bar{\psi}) = \bar{\psi}(x)(i\gamma^\mu D_\mu - m)\psi(x), \tag{3.41}$$

where

$$D_\mu = \partial_\mu - g_s A_\mu. \tag{3.42}$$

The $SU(3)$ gauge filed A_μ is defined as

$$A^\mu = \vec{A}_\mu \cdot \vec{\rho} = \sum A^i_\mu \rho_i. \tag{3.43}$$

The wave function $\psi(x)$ is triplet under $SU(3)$, $\psi = \begin{pmatrix} \psi_1 \\ \psi_2 \\ \psi_3 \end{pmatrix}$, with the following transformation:

$$\psi(x) \to \psi'(x) = U\psi(x), \qquad \bar{\psi}(x) \to \bar{\psi}(x)' = \bar{\psi}(x)U^\dagger, \tag{3.44}$$

where U is a matrix in $SU(3)$ given in terms of the Gell-Mann representation of Eq. (3.36). Then, the invariant Lagrangian takes the form

$$\mathscr{L}(\psi,\bar{\psi}) = \bar{\psi}(x)(i\gamma^\mu\partial_\mu - m)\psi(x) - \frac{1}{2}ig_s A^i_\mu \bar{\psi}\gamma^\mu\lambda_i\psi. \tag{3.45}$$

So the field interactions are given by

$$\mathscr{L}_I = \frac{1}{2} i g_s A^i_\mu \bar{\psi} \gamma^\mu \lambda_i \psi. \tag{3.46}$$

As in the $U(1)$ and $SU(2)$ gauge symmetries, one must add the kinetic term of the gauge fields. Thus, a field stress tensor is defined as follows:

$$F_{\mu\nu} = \partial_\mu A_\nu - \partial_\nu A_\mu + g_s[A_\mu, A_\nu], \tag{3.47}$$

$$F^i_{\mu\nu} = \partial_\mu A^i_\nu - \partial_\nu A^i_\mu + g_s \epsilon_{abc} A^b_\mu A^c_\nu, \tag{3.48}$$

where ϵ_{abc} is the fully antisymmetric Levi-Civita tensor in three dimensions, which leads to the following kinematic interaction term

$$-\frac{1}{2}\text{Tr}(F_{\mu\nu}, F^{\mu\nu}) = -\frac{1}{4} F^a_{\mu\nu} F^{a\mu\nu}. \tag{3.49}$$

Therefore, the total Lagrangian, which is invariant under $SU(3)$, is given by

$$\mathscr{L}_0(\psi, \bar{\psi}) = \bar{\psi}(x)(i\gamma^\mu \partial_\mu - m)\psi(x) - \frac{1}{2} i g_s A^i_\mu \bar{\psi} \gamma^\mu \lambda_i \psi - \frac{1}{4} F^a_{\mu\nu} F^{a\mu\nu}. \tag{3.50}$$

Here too, the mass of the gauge field must vanish because of the gauge symmetry.

The physical significance of the groups discussed in this chapter will be laid out in the subsequent ones. The three gauge fields of $SU(2)$ will be considered to account for the weak vector boson fields $W^\pm$ and Z while the 8 gauge fields of $SU(3)$ will correspond to the strong interaction force carriers, which are called *gluons*.

CHAPTER 4

Theory of EW Interactions

There were several attempts to construct a gauge theory of weak interactions. As shown in Chapter 2, that weak processes are mediated by two $W^{\pm}$ bosons; therefore the first trial was based on the $SU(2)$ weak isospin group, which includes three gauge bosons. The hope was that two of these gauge bosons corresponded to $W^{\pm}$ and the third one would be the EM gauge boson (photon). However, as we will show below, this trial was not successful. In 1961, Glashow proposed that both weak and EM interactions can be implemented in the gauge group $SU(2) \times U(1)$ [27], where $U(1)$ was associated with the leptonic hypercharge (Y).

4.1 $SU(2)_L$ GAUGE SYMMETRY

We start by assuming that left-handed leptons, Ψ_{lL}, and left-handed quarks, Ψ_{qL}, transform like doublets of the isotopic group $SU(2)_L$, namely,

$$\Psi_{eL} = \begin{pmatrix} \nu'_{eL} \\ e'_L \end{pmatrix}, \quad \psi_{\mu L} = \begin{pmatrix} \nu'_{\mu L} \\ \mu'_L \end{pmatrix}, \quad \Psi_{\tau L} = \begin{pmatrix} \nu'_{\tau L} \\ \tau'_L \end{pmatrix}, \tag{4.1}$$

$$\Psi_{1L} = \begin{pmatrix} u'_L \\ d'_L \end{pmatrix}, \quad \Psi_{2L} = \begin{pmatrix} c'_L \\ s'_L \end{pmatrix}, \quad \Psi_{3L} = \begin{pmatrix} t'_L \\ b'_L \end{pmatrix}. \tag{4.2}$$

DOI: 10.1201/9780429443015-4

In contrast, l_R, q_R transform like singlets. As usual, the free Lagrangian of massless particles is given by

$$\mathscr{L}_0 = \bar{\Psi} i\gamma^\alpha \partial_\alpha \Psi. \tag{4.3}$$

Therefore $\mathscr{L}_0$ takes the form

$$\begin{aligned}\mathscr{L}_0 &= \sum \bar{\nu}'_{lL} i\gamma^\alpha \partial_\alpha \nu'_{lL} + \sum \bar{l}_L i\gamma^\alpha \partial_\alpha l'_L + \sum \bar{l}_R i\gamma^\alpha \partial_\alpha l'_R \\ &+ \sum \bar{q}'_L i\gamma^\alpha \partial_\alpha q'_L + \sum \bar{q}'_R i\gamma^\alpha \partial_\alpha q'_R,\end{aligned} \tag{4.4}$$

which can be written as

$$\begin{aligned}\mathscr{L}_0 &= \sum \bar{\Psi}_{lL} i\gamma^\alpha \partial_\alpha \Psi_{lL} + \sum \bar{\Psi}_{qL} i\gamma^\alpha \partial_\alpha \Psi_{qL} \\ &+ \sum \bar{l}'_R i\gamma^\alpha \partial_\alpha l'_R + \sum \bar{q}'_R i\gamma^\alpha \partial_\alpha q'_R.\end{aligned} \tag{4.5}$$

This free Lagrangian is invariant under the considered isotopic (global) transformation:

$$\begin{aligned}\Psi'_L &= e^{i\frac{1}{2}\vec{\tau}.\vec{A}} \Psi_L, \\ l'_R, q'_R &= l_R, q_R.\end{aligned} \tag{4.6}$$

As previously seen, the Lagrangian could be invariant under local $SU(2)_L$ symmetry if normal derivatives of left-handed fields are changed to covariant derivatives, as follows:

$$\partial_\alpha \Psi_{lL} \to \left(\partial_\alpha + ig\frac{1}{2}\vec{\tau}\cdot\vec{A}\right)\Psi_{lL} \tag{4.7}$$

and

$$\partial_\alpha \Psi_{qL} \to \left(\partial_\alpha + ig\frac{1}{2}\vec{\tau}\cdot\vec{A}\right)\Psi_{qL}, \tag{4.8}$$

where the gauge field A^i_α is a vector field triplet. In this case, the interaction Lagrangian is given by $\mathscr{L}_I = -g\vec{J}_\alpha \vec{A}^\alpha$, where g is the dimensionless gauge coupling constant and

$$J_\alpha = \sum \bar{\Psi}_{lL}\gamma_\alpha \frac{1}{2}\vec{\tau}\Psi_{lL} + \sum \bar{\Psi}_{iL}\gamma_\alpha \frac{1}{2}\vec{\tau}\Psi_{iL}. \tag{4.9}$$

The above interaction Lagrangian can also be rewritten as

$$\mathscr{L}_I = \frac{-g}{2\sqrt{2}} j^{\alpha} W_{\alpha} + h.c - g j^3_{\alpha} A^{3\alpha}, \tag{4.10}$$

where $W_{\alpha} = (A^1_{\alpha} - iA^2_{\alpha})/\sqrt{2}$ is the field of the charged vector particles and the charged current j_{α} is given by

$$\begin{aligned} j_{\alpha} &= 2j^{1+2i}_{\alpha} = 2(j^1_{\alpha} + ij^2_{\alpha}) \\ &= 2\sum \bar{\nu}'_{lL} \gamma_{\alpha} l'_L + 2(\bar{u}'_L \gamma_{\alpha} d'_L + \bar{c}'_L \gamma_{\alpha} s'_L + ...). \end{aligned} \tag{4.11}$$

The neutral current j^3_{α} is given by

$$j^3_{\alpha} = \frac{1}{2}\sum \bar{\nu}'_{lL}\gamma_{\alpha}\nu'_{lL} - \frac{1}{2}\sum \bar{l}'_L\gamma_{\alpha}l'_L + \frac{1}{2}\sum \bar{u}'_{iL}\gamma_{\alpha}u'_{iL} - \frac{1}{2}\sum \bar{d}'_{iL}\gamma_{\alpha}d'_{iL}. \tag{4.12}$$

The first term of the Lagrangian in Eq. (4.10) describes the interaction between the charged intermediate gauge bosons and the charged weak current, while the last term in Eq. (4.10) describes the interaction with neutral vector bosons. It is clear that this term cannot be the Lagrangian of the EM interaction, for the following reasons. (i) Here, j^3_{α} has no right-handed components while the EM current j^{EM}_{α} of the electron, for instance, has both left- and right-handed components, *i.e.*, $j^{\mathrm{EM}}_{\alpha} = -\bar{e}\gamma_{\alpha}e = -\bar{e}_L\gamma_{\alpha}e_L - \bar{e}_R\gamma_{\alpha}e_R$. (ii) The charge of the neutrino, which couples to A^3_{α}, is not zero, but opposite to the one of the electron, as $j^3_{\alpha} = \frac{1}{2}\bar{\nu}_e\gamma_{\alpha}\nu_e - \frac{1}{2}\bar{e}_L\gamma_{\alpha}e_L$, while it should be zero in the EM interaction. (iii) Contrary to j^{EM}_{α}, the neutral current j^3_{α} maximally violates parity.

Thus, one can conclude that the gauge symmetry $SU(2)_L$ should be enlarged by a new $U(1)$ symmetry, called $U(1)_Y$. This symmetry should be independent of the $SU(2)_L$ group and thus its generator should commute with the $SU(2)_L$ generators. Also, we should obtain the correct form of j^{EM}_{μ} for a fermion ψ with charge Q as $j^{\mathrm{EM}}_{\mu} = \bar{\psi}\gamma_{\mu}Q\psi$. Moreover, the new symmetry will be chosen such that the electric charge Q can be given as a linear combination of the generator T^3 of $SU(2)_L$ and the generator of the new $U(1)_Y$ group. It is interesting to note the relation

$$Q - T^3 = \int d^3x \left(-\frac{1}{2}\nu^+_e \nu_e - \frac{1}{2}e^+_L e_L - e^+_R e_R \right), \tag{4.13}$$

Particle	Q	T^3	Y
ν_ℓ	0	1/2	−1
ℓ_L	−1	−1/2	−1
ℓ_R	−1	0	−2
u_L	2/3	1/2	1/3
d_L	−1/3	−1/2	1/3
u_R	2/3	0	4/3
d_R	−1/3	0	−2/3

Table 4.1 Weak hypercharge of quarks and leptons.

which shows that the elements in the $SU(2)_L$ doublet have the same eigenvalue $-1/2$ and the eigenvalue of e_R is -1. Furthermore, $Q - T^3$ commutes with T^i $(i = 1, 2, 3)$ of $SU(2)_L$, so that Q and T^i can be simultaneous symmetries of the model. Therefore, it is reasonable to define the new $U(1)_Y$ such that its eigenvalue Y is obtained as [28, 29]

$$Y/2 = Q - T^3, \tag{4.14}$$

or

$$Q = T^3 + Y/2. \tag{4.15}$$

The eigenvalue Y is called weak hypercharge and its values for quarks and leptons are given in Tab. 4.1.

4.2 $SU(2)_L \times U(1)_Y$ GAUGE SYMMETRY

Let us now extend the group of transformations under which the Lagrangian of the system under consideration must be invariant. Namely, let us impose the condition that the Lagrangian be invariant with respect to the direct product of the local $SU(2)_L$ and $U(1)_Y$ groups [30, 31]. In this case, the left-handed leptons and quarks interact both with the triplet of gauge fields $A_\alpha(x)$ and the additional gauge field $B_\alpha(x)$ connected with the local hypercharge group, *i.e.*,

$$\partial_\alpha \Psi_{lL} \rightarrow \left(\partial_\alpha + ig\frac{1}{2}\vec{\tau}.\vec{A} + \underbrace{ig'\frac{1}{2}y_L}_{\text{constant}} B_\alpha \right) \Psi_{lL}, \tag{4.16}$$

$$\partial_\alpha \Psi_{qL} \rightarrow \left(\partial_\alpha + ig\frac{1}{2}\vec{\tau}.\vec{A} + ig'\frac{1}{2}y_L^q B_\alpha \right) \Psi_{qL}\,, \tag{4.17}$$

while the right-handed leptons and quarks interact only with the gauge field $B_\alpha(x)$ of $U(1)_Y$. Thus,

$$\partial_\alpha l_R^{'} \rightarrow \left(\partial_\alpha + ig^{'}\frac{1}{2}y_R B_\alpha\right) l_R^{'}, \tag{4.18}$$

$$\partial_\alpha q_R^{'} \rightarrow \left(\partial_\alpha + ig^{'}\frac{1}{2}y_R^q B_\alpha\right) q_R^{'}. \tag{4.19}$$

We then arrive at the following Lagrangian of the interactions with the gauge fields A_α and B_α:

$$\mathscr{L}_I = -\frac{g}{2\sqrt{2}} J_\alpha W^\alpha + h.c - gJ_\alpha^3 A^{3\alpha} - g^{'}\frac{1}{2}J_\alpha^y B^\alpha. \tag{4.20}$$

The charged interactions (first term in the above Lagrangian) for left-handed leptons (and similarly for left-handed quarks) can be written as

$$\mathcal{L}_I^{\pm} = -\frac{g}{2\sqrt{2}}\left[\bar{\nu}\gamma^\mu(1-\gamma_5)\ell W^+ + \bar{\ell}\gamma^\mu(1-\gamma_5)\nu W_\mu^-\right], \tag{4.21}$$

which reproduces exactly the $V - A$ structure of the weak charged current discussed in Chapter 2. Considering Fermi theory as an effective low energy limit of the above Lagrangian, one obtains

$$M_W = \frac{g}{\sqrt{2}} = \left(\frac{M_W^2 G_F}{\sqrt{2}}\right)^{1/2}, \tag{4.22}$$

which indicates that the mass of the W^+ boson is of order 80 GeV. In addition, J_α^y is given by

$$J_\alpha^y = \sum y_L \bar{\Psi}_{lL}\gamma_\alpha \Psi_{lL} + \sum y_L^q \bar{\Psi}_{kL}\gamma_\alpha \Psi_{kL} + \sum \bar{l}_R^{'}\gamma_\alpha l_R^{'} + \sum y_R \bar{q}_R^{'}\gamma^\alpha q_R^{'}. \tag{4.23}$$

Hence, for leptons only, one finds

$$\frac{1}{2}J_\alpha^y = -\frac{1}{2}\sum \bar{\Psi}_{lL}\gamma_\alpha \Psi_{lL} - \sum \bar{l}_R^{'}\gamma_\alpha l_R^{'} \tag{4.24}$$

and

$$\begin{aligned}
\frac{1}{2}J_\alpha^y + J_\alpha^3 &= -\frac{1}{2}\sum \bar{\nu}_{lL}^{'}\gamma_\alpha \nu_{lL} - \frac{1}{2}\sum \bar{l}_L^{'}\gamma_\alpha l_L^{'} + \frac{1}{2}\sum \bar{\nu}_{lL}^{'}\gamma_\alpha \nu_{lL}^{'} \\
&- \frac{1}{2}\sum \bar{l}_L^{'}\gamma_\alpha l_L - \sum \bar{l}_R^{'}\gamma_\alpha l_R^{'} \\
&= -\sum \bar{l}_L^{'}\gamma_\alpha l_L^{'} - \sum \bar{l}_R^{'}\gamma_\alpha l_R^{'} \\
&= -\sum \bar{l}^{'}\gamma_\alpha l^{'} = J_\alpha^{\mathrm{EM}},
\end{aligned} \tag{4.25}$$

i.e.,

$$J_\alpha^{\mathrm{EM}} = J_\alpha^3 + \frac{1}{2} J_\alpha^y. \tag{4.26}$$

Same results can be obtained for quarks as well. From this equation, it is possible to single out the Lagrangian of the EM interaction as follows:

$$\begin{aligned}
\mathscr{L}_I^0 &= -g j_\alpha^3 A^{3\alpha} - g^{'} (j_\alpha^{\mathrm{EM}} - j_\alpha^3) B^\alpha \\
&= -\sqrt{g^2 + g^{'2}} j_\alpha^3 \left(\frac{g A^{3\alpha}}{\sqrt{g^2 + g^{'2}}} - \frac{g^{'} B^\alpha}{\sqrt{g^2 + g^{'2}}} \right) - g^{'} g_\alpha^{\mathrm{EM}} B^\alpha.
\end{aligned} \tag{4.27}$$

If one adopts the following field redefinitions

$$Z^\alpha = \frac{g A^{3\alpha}}{\sqrt{g^2 + g^{'2}}} - \frac{g^{'} B^\alpha}{\sqrt{g^2 + g^{'2}}}, \tag{4.28}$$

$$A^\alpha = \frac{g^{'} A^{3\alpha}}{\sqrt{g^2 + g^{'2}}} + \frac{g B^\alpha}{\sqrt{g^2 + g^{'2}}}, \tag{4.29}$$

one finds

$$\begin{aligned}
\mathscr{L}_I^0 &= -\sqrt{g^2 + g'^2} j_\alpha^3 Z^\alpha - \frac{g g^{'}}{\sqrt{{}^2 + g^{'2}}} j_\alpha^{\mathrm{EM}} A^\alpha + \frac{g^{'2}}{\sqrt{g^2 + g^{'2}}} j_\alpha^{\mathrm{EM}} Z^\alpha \\
&= -\frac{1}{2} \sqrt{g^2 + g'^2} j_\alpha^z Z^\alpha - \frac{g g^{'}}{\sqrt{g^2 + g^{'2}}} j_\alpha^{\mathrm{EM}} A^\alpha,
\end{aligned} \tag{4.30}$$

where

$$j_\alpha^z = 2 j_\alpha^3 - \frac{2 g'^2}{g^2 + g'^2} j_\alpha^{\mathrm{EM}}. \tag{4.31}$$

Let us introduce the Weinberg (or weak) angle as follows:

$$\tan \theta_W = \frac{g^{'}}{g}, \tag{4.32}$$

so that

$$\sqrt{g^2 + g^{'2}} = \frac{g}{\cos \theta_W}, \qquad \frac{g g^{'}}{\sqrt{g^2 + g'^2}} = g \sin \theta_W.$$

Hence,

$$\mathscr{L}_I^0 = -\frac{g}{2\cos\theta_W} j_\alpha^z Z^\theta - g\sin\theta_W j_\alpha^{\mathrm{EM}} A^\alpha, \tag{4.33}$$

where

$$j_\alpha^z = 2j_\alpha^3 - 2\sin^2\theta_W j_\alpha^{\mathrm{EM}}. \tag{4.34}$$

We also require that

$$g\sin\theta_W = e, \tag{4.35}$$

i.e.,

$$\frac{1}{e^2} = \frac{1}{g^2} + \frac{1}{g'^2}. \tag{4.36}$$

Thus the unification of weak and EM interactions on the basis of gauge invariance is possible. We have constructed the Lagrangian of EW interactions taking as a basis gauge invariance. Such a theory is renormalisable. However, it is in disagreement with experimental data, which confirmed the non-vanishing masses of the involved particles (both weak intermediate bosons and fermions, except neutrinos). In order to construct a renormalisable theory of EW interactions for physical massive particles, we must spontaneously break the gauge symmetry as we will show in the next chapter.

CHAPTER 5

SSB and Higgs Mechanism

In a previous chapter, a gauge invariant and renormalisable unified theory of weak and EM interactions, henceforth EW, was obtained. However, all leptons and gauge bosons had to have zero masses. We have also seen that, if we put masses by hand, the Lagrangian density will not remain gauge invariant and the theory will be non-renormalisable. In order to obtain a renormalisable theory, it is essential to introduce the gauge boson masses (where required) via a mechanism which retains the gauge invariance of the Lagrangian density. This mechanism is called SSB.

5.1 SYMMETRY BREAKING

There are many types of symmetries in our Universe, such as space-time symmetries (*e.g.*, Lorentz symmetry, P, time reversal, etc.) and internal symmetries, which do not depend on space-time (*e.g.*, isospin, flavour and colour symmetries, etc.). Some symmetries are called exact symmetries and some others are called approximate symmetries. We focus on symmetry breaking in the physical world. This breaking can be done in two ways. In the first one we put a symmetry breaking term in Lagrangian by hand as follows:

$$\mathscr{L} = \mathscr{L}_{\text{symmetric}} + \mathscr{L}_{\text{breaking}}. \tag{5.1}$$

DOI: 10.1201/9780429443015-5

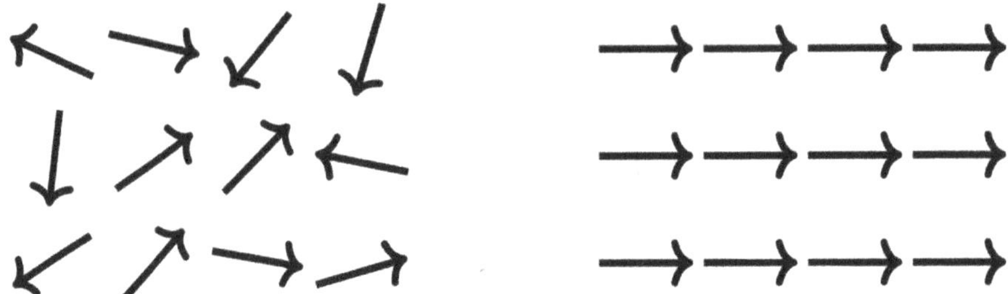

Figure 5.1 (Left) Arbitrary spin configuration of a ferromagnet at high temperature. (Right) At low temperatures, the system has definite spin.

This approach to symmetry breaking is useful and meaningful when the symmetry breaking term is small and thus the theory will be perturbative but this term will lead to a non-renormalisable theory and our theory will be irrelevant. The second way is by exploiting a hidden or SSB. In this scenario, one considers a system with a Lagrangian $\mathscr{L}$ that is invariant under a symmetry transformation. Then by considering the lowest state of the system (ground state), one finds two cases. First, the ground state is unique (not degenerate) and hence it is invariant under the symmetry transformation of $\mathscr{L}$. Second, the ground state may not be unique (degenerate), *i.e.*, there is no unique eigenstate to represent the ground state. If we arbitrary select one of the degenerate states as the ground state, then the ground state will not be invariant under the same symmetry transformation of $\mathscr{L}$. The mechanism of selecting one ground state from the degenerate ones is known as spontaneous (*i.e.*, this means that the symmetry is not broken explicitly) symmetry breaking.

There are many examples of SSB. The example of a ferromagnet is widely used in contexts of SSB. At high temperatures, greater than the Curie temperature of the material, the spins point in arbitrary directions as in Fig. 5.1 (left) and net magnetisation of the system is equal to zero. In this case, the spin system is invariant under rotations due to having disordered spins. When the temperature of the system becomes lower than the Curie one, the spins begin to align and there is a net magnetisation with a specific direction as shown in Fig. 5.1 (right). This alignment breaks rotational symmetry of such a spin system spontaneously.

5.2 SPONTANEOUS SYMMETRY BREAKING OF A GLOBAL $U(1)$

Let us begin with a Lagrangian which describes a complex scalar field ϕ, which can be written as

$$\mathscr{L} = \partial_\mu \phi^*(x)\partial^\mu \phi(x) - V(\phi), \tag{5.2}$$

with

$$\phi(x) = \frac{1}{\sqrt{2}}(\phi_1 + i\phi_2) \tag{5.3}$$

and the following form of the scalar potential:

$$V(\phi) = \mu^2|\phi(x)|^2 + \lambda|\phi(x)|^4. \tag{5.4}$$

The Lagrangian density in Eq. (5.2) is invariant under the $U(1)$ global gauge transformations

$$\phi(x) \quad \rightarrow \quad \phi'(x) = e^{i\alpha}\phi(x), \tag{5.5}$$

$$\phi^*(x) \quad \rightarrow \quad \phi^{*\prime}(x) = e^{-i\alpha}\phi^*(x). \tag{5.6}$$

In terms of ϕ_1 and ϕ_2, the scalar potential $V(\phi)$ takes the form

$$V(\phi_1, \phi_2) = \frac{\mu^2}{2}(\phi_1^2 + \phi_2^2) + \frac{\lambda}{4}(\phi_1^2 + \phi_2^2)^4. \tag{5.7}$$

It is worth noting that the potential $V(\phi)$ must be at most 4th order in the fields in order to ensure renormalisability of the theory and the coupling λ must be positive to ensure that theory is bounded from below. By minimising $V(\phi)$, one finds two possible minima, depending on the sign of μ^2.

The first minimum corresponds to $\mu^2 > 0$. In this case, $V(\phi)$ has an absolute minimum at $\phi = 0$ and the above mentioned global symmetry remains exact. In this regard, we have a stable equilibrium (*i.e.*, a minimum) at the point $\phi = 0$ and fluctuations around this point give rise to a spin-0 particle with mass μ and a perturbative self-interacting terms. In case $\mu^2 < 0$, the potential $V(\phi)$ has a local maximum at $\phi = 0$ and a whole circle of absolute minima at

$$\phi_0 = \sqrt{\frac{-\mu^2}{2\lambda}}e^{i\theta}, \qquad 0 < \theta < 2\pi, \tag{5.8}$$

where θ defines a direction in the complex ϕ plane. The potential energy density in this phase is displayed in Fig. 5.2. As it can be seen, the vacuum

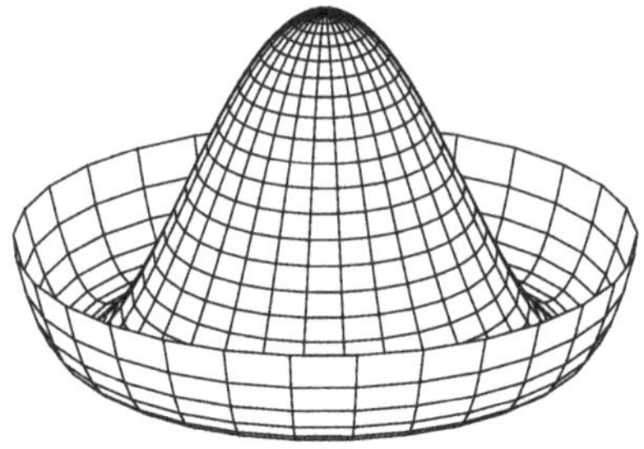

Figure 5.2 Mexican hat potential.

state is not unique and SSB can occur by choosing a specific direction of θ. If the $U(1)$ symmetry is broken spontaneously and a specific direction with $\theta = 0$ is considered, then the VEV of ϕ is given by

$$\langle 0|\phi|0\rangle = \sqrt{\frac{-\mu^2}{2\lambda}} = \frac{v}{\sqrt{2}} > 0. \tag{5.9}$$

It is important to note that now $\phi = 0$ is an unstable point and fluctuations around this point would give rise to a spin-0 particle with imaginary mass. Also, we cannot do any perturbative calculations around an unstable solution. Therefore, we rewrite $\phi(x)$ in terms of two real fields $h(x)$ and $\eta(x)$ as follows:

$$\phi(x) = \frac{1}{\sqrt{2}}(v + h(x) + i\eta(x)). \tag{5.10}$$

Now, the VEVs of the shifted scalar fields are given by $\langle 0|h|0\rangle = 0$ and $\langle 0|\eta|0\rangle = 0$. Also, Eq. (5.10) will be given by

$$\begin{aligned} \mathscr{L} &= \frac{1}{2}\partial^\mu h(x)\partial_\mu h(x) + \frac{1}{2}\partial^\mu \eta(x)\partial_\mu \eta(x) - \lambda v h^2(x) \\ &- \lambda v h(x)\left[h^2(x) + \eta^2(x)\right] - \frac{\lambda}{4}\left[h^2(x) + \eta^2(x)\right]^2. \end{aligned} \tag{5.11}$$

There is an additional constant term in Eq. (5.11), which is irrelevant in our analysis hence it is omitted. It is clear that the field $h(x)$ is a boson with a real and positive mass, which is given by $\sqrt{2\lambda v}$, while the field $\eta(x)$ is a massless boson. The latter is called a Goldstone boson [32].

We can then write the complex scalar field $\phi(x)$ in polar coordinates by introducing two real scalar fields $\rho(x)$ and $\eta(x)$,

$$\phi = \frac{1}{\sqrt{2}}\rho\, e^{\frac{i\eta}{v}}, \tag{5.12}$$

where v is constant and given by $v = \sqrt{\frac{\mu^2}{\lambda}}$. Then, by substituting into Eq. (5.2), we have

$$\partial_\mu \phi = \frac{1}{\sqrt{2}} e^{\frac{i\eta}{v}} \left(\partial_\mu \rho + \frac{i}{v} \rho \partial_\mu \eta \right). \tag{5.13}$$

We further have

$$\mathscr{L} = \frac{1}{2}(\partial_\mu \rho)^2 + \frac{1}{2v^2}\rho^2(\partial_\mu \eta)^2 - \frac{\mu^2}{2}\rho^2 - \frac{\lambda}{4}\rho^4. \tag{5.14}$$

As mentioned, by taking $\mu^2 < 0$, the $U(1)$ global symmetry is broken by selecting one specific vacuum state from the degenerate ones and by inducing a deviation through the real scalar field $h(x)$ in the radial direction, so that one eventually finds a massive particle as shown by the following Lagrangian:

$$\mathscr{L} = \frac{1}{2}(\partial_\mu h)^2 + \frac{1}{2}(\partial_\mu \eta)^2 + \frac{1}{v}h(\partial_\mu \eta)^2 + \frac{1}{2v^2}h^2(\partial_\mu \eta)^2 - V(\rho^2), \tag{5.15}$$

where

$$V(\rho^2) = \frac{1}{2}(2\mu^2)h^2 + \lambda v h^3 + \frac{\lambda}{4}h^4 - \frac{1}{4}\mu^2 v^2. \tag{5.16}$$

From this equation, as there is no mass term for $\eta(x)$, it is then clear that this is the aforementioned Goldstone boson which is generated as a result of spontaneously breaking the $U(1)$ symmetry (which has one generator) by an unstable vacuum and we have also a massive $h(x)$ field with mass $\sqrt{2\mu^2}$.

5.3 SPONTANEOUS SYMMETRY BREAKING OF A GLOBAL $SU(2)$

We now consider the SSB of a non-Abelian global $SU(2)$ symmetry. The scalar field ϕ which belongs to the fundamental representation of $SU(2)$ can be written as

$$\phi = \begin{pmatrix} \phi_1 \\ \phi_2 \end{pmatrix},$$

where ϕ_1 and ϕ_2 are complex scalar fields. The $SU(2)$ global gauge invariant Lagrangian is given by

$$\mathscr{L} = \partial_\mu \phi^\dagger \partial^\mu \phi - \mu^2 \phi^\dagger \phi - \lambda(\phi^\dagger \phi)^2, \tag{5.17}$$

with $\mu^2 < 0$. Applying the above procedure used in $U(1)$ spontaneous breaking and writing ϕ in polar coordinates, upon introducing 4 real scalar fields $h(x)$ and $\eta^i(x)$ $(i = 1, 2, 3)$, we obtain ϕ as

$$\phi = \frac{1}{\sqrt{2}} e^{\frac{i\tau^i \eta^i}{2v}} \begin{pmatrix} 0 \\ v + h \end{pmatrix}, \tag{5.18}$$

where the VEV is given by $v = (\phi_1^0)^2 + (\phi_2^0)^2 = \frac{\mu^2}{\lambda}$. Then

$$\partial_\mu \phi = \frac{1}{\sqrt{2}} e^{\frac{i\tau^i \eta^i}{2v}} \left[\begin{pmatrix} 0 \\ \partial_\mu h \end{pmatrix} + \frac{i}{v}\frac{\tau^i}{2} \partial_\mu \eta^i \begin{pmatrix} 0 \\ v + h \end{pmatrix} \right]. \tag{5.19}$$

By substituting into Eq. (5.17), one gets

$$\begin{aligned} \mathscr{L} &= \frac{1}{2} \left[\begin{pmatrix} 0 & \partial_\mu h \end{pmatrix} - \frac{i}{v} \partial_\mu \eta^i \begin{pmatrix} 0 & v + h \end{pmatrix} \frac{\tau^i}{2} \right] \\ &\times \left[\begin{pmatrix} 0 \\ \partial^\mu h \end{pmatrix} + \frac{i}{v}\frac{\tau^j}{2} \partial_\mu \eta^j \begin{pmatrix} 0 \\ v + h \end{pmatrix} \right] - V(h), \end{aligned} \tag{5.20}$$

which leads to

$$\mathscr{L} = \frac{1}{2} \partial_\mu h \partial^\mu h + \frac{1}{8v^2} \partial_\mu \eta^i \partial^\mu \eta^i (v + h)^2 - V(h), \tag{5.21}$$

with $V(h)$ given by

$$V(h) = \frac{1}{2}(2\mu^2)h^2 + \lambda v h^3 + \frac{\lambda}{4} h^4 - \frac{\mu^2 v^2}{4}. \tag{5.22}$$

From Eqs. (5.21) and (5.22), one sees that the original $SU(2)$ symmetry is broken and three Goldstone bosons, corresponding to now three broken $SU(2)$ generators, appear, in addition to a massive particle with mass equal $\sqrt{2\mu^2}$.

5.4 THE GOLDSTONE THEOREM

The above conclusion can be generalised to cases with larger symmetries. Let us consider a complex scalar field which belongs to the fundamental representation of a $SU(n)$ group:

$$\phi = \begin{pmatrix} \phi_1 \\ \phi_2 \\ \vdots \\ \phi_n \end{pmatrix}. \tag{5.23}$$

The $SU(n)$ group has $N = n^2 - 1$ generators G^i $(i = 1, ..., N)$. In addition, ϕ transforms under $SU(n)$ as follows:

$$\phi \to \phi' = e^{-i\alpha^i G^i}\phi. \tag{5.24}$$

Thus, the corresponding infinitesimal transformation is given by

$$\delta\phi = -i\alpha^i G^i \phi, \tag{5.25}$$

where G^i is $n \times n$ matrix. Suppose that the vacuum is not unique and it is symmetrically degenerate. Then we break the symmetry spontaneously and choose one particular vacuum,

$$\phi_0 = \begin{pmatrix} \phi_1 \\ \phi_2 \\ \vdots \\ \phi_n \end{pmatrix}_0. \tag{5.26}$$

The vacuum state in Eq. (5.26) is no longer invariant under the infinitesimal action of the generators (*i.e.*, $\delta\phi_0 = -i\alpha^i G^i \phi_0$). Now, suppose that there are M $(\leq N)$ generators G^i $(i = 1, ..., M)$ breaking the vacuum invariance while the others $G^i (i = M+1, ..., N)$ leave the vacuum invariant ($G^i\phi_0 = 0$). If one studies the fluctuations around the stable minimum ϕ_0 (*i.e.*, $\phi = \phi_0 + \phi^{'}$) and expands the potential around this point to get physical spectra, then one gets

$$V = V_0 + \sum_{a=1}^{n} \left(\frac{\partial V}{\partial \phi_a}\right)_0 \phi_a^{'} + \frac{1}{2}\sum_{a,b=1}^{n} \left(\frac{\partial^2 V}{\partial \phi_a \phi_b}\right)_0 \phi_a^{'}\phi_b^{'} + ... \ , \tag{5.27}$$

where

$$\left(\frac{\partial V}{\partial \phi_a}\right)_0 = 0 \quad \text{and} \quad m^2_{ab} = \left(\frac{\partial^2 V}{\partial \phi_a \phi_b}\right)_0. \tag{5.28}$$

The potential should be invariant under $SU(n)$ transformations, *i.e.*, $\delta V = \frac{\partial V}{\partial \phi_a}\delta\phi_a$. By using Eq. (5.24), one finds

$$\sum_{a,b} \frac{\partial V}{\partial \phi_a} G^i_{ab}\phi_b = 0; \qquad i = 1, ..., N. \tag{5.29}$$

By differentiating this equation with respect to ϕ_c, we get

$$\sum_{a,b} \frac{\partial^2 V}{\partial \phi_c \phi_a} G^i_{ab}\phi_b + \sum_{a} \frac{\partial V}{\partial \phi_a} G^i_{ac} = 0. \tag{5.30}$$

At the vacuum point $\phi = \phi_0$, so the previous equation becomes

$$\sum_{a,b} m^2_{ca} G^i_{ab}(\phi_b)_0 = 0, \quad for \ \ i = 1, ..., N. \tag{5.31}$$

For the generators which break the symmetry (*i.e.*, $G^i\phi_0 \neq 0$), there are M number of particles with zero mass eigenvalue (*i.e.*, $m^2_{ca} = 0$). In other words, for each generator breaking the symmetry at the vacuum, there exists a massless Goldstone boson. This is known as the Goldstone theorem [32].

5.5 HIGGS MECHANISM IN GAUGE THEORIES

In the above sections, we have considered SSB of global symmetries. We now extend our analysis to local (*i.e.*, gauge) symmetries. As advocated in previous chapters, to make a Lagrangian invariant under a local symmetry, one must replace partial with covariant derivatives. As we will see, breaking the local symmetry spontaneously by choosing one particular vacuum from degenerate ones, the corresponding gauge fields will acquire masses and all massless

Goldstone bosons will disappear. This mechanism is known as the Higgs mechanism [33–36]. We will start analysing this with an Abelian $U(1)$ gauge symmetry.

5.5.1 $U(1)$ Gauge Symmetry

The Lagrangian density in Eq. (5.2) can be generalised to be $U(1)$ gauge invariant by introducing a gauge field $A_\mu(x)$ and replacing the ordinary derivative by a covariant derivative:

$$D_\mu \phi = (\partial_\mu - ieA_\mu)\phi.$$

Then, the Lagrangian density becomes

$$\mathscr{L} = (D^\mu \phi^*)(D_\mu \phi) - \mu^2 |\phi|^2 - \lambda |\phi|^4 - \frac{1}{4} F_{\mu\nu} F^{\mu\nu}. \tag{5.32}$$

The Lagrangian density in Eq. (5.32) is invariant under $U(1)$ gauge transformations, *i.e.*,

$$\phi \to \phi' = e^{-i\alpha(x)}\phi, \tag{5.33}$$

$$A_\mu \to A'_\mu - \frac{1}{e}\partial_\mu \alpha(x). \tag{5.34}$$

Again, we take $\lambda > 0$ to ensure that the system will be bounded from below and, again, two situations arise depending on the sign of μ^2. For $\mu^2 > 0$ the state of lowest energy (*i.e.*, the vacuum state) corresponding to $\phi(x) = 0$ is unique and thus the symmetry remains exact. For $\mu^2 < 0$, the vacuum state is not unique and SSB occurs. To ensure Lorentz invariance, we must have $\langle 0|A_\mu|0\rangle = 0$ thus again obtaining a circle of local minima as in Eq. (5.8) with the point $\phi(x) = 0$ representing a local maximum. We now choose one particular direction θ to represent the vacuum state as we did in Eq. (5.9) and introduce two real fields $h(x)$ and $\eta(x)$ as in Eq. (5.10). Therefore, the Lagrangian density in Eq. (5.32) becomes

$$\mathscr{L} = \frac{1}{2}(\partial^\mu h)(\partial_\mu h) - \frac{1}{2}(2\lambda v^2)h^2 - \frac{1}{4}F_{\mu\nu}F^{\mu\nu} + \frac{1}{2}(ev)^2 A_\mu A^\mu$$

$$+\frac{1}{2}(\partial^\mu \eta)(\partial_\mu \eta) - evA^\mu \partial_\mu \eta + \mathscr{L}_I, \tag{5.35}$$

where $\mathscr{L}_I$ represents interaction terms, which are cubic and quartic in the fields. The first two terms in Eq. (5.35) describe a real Klein-Gordon field, which represents a neutral spin-0 particle, with mass $\sqrt{2\lambda v^2}$. However, the product term $A^\mu \partial_\mu \eta$ indicates that the two fields A^μ and η are not in canonical form. This ambiguity can be seen by counting the number of degrees of freedom of the Lagrangian densities in Eqs. (5.32) and (5.35). In Eq. (5.32), there are four degrees of freedom: two from the complex scalar field $\phi(x)$ and two from the massless vector field $A_\mu(x)$ (two transverse polarisation states). In contrast, in Eq. (5.35), there are five degrees of freedom: two from the real scalar fields $h(x)$ and $\eta(x)$ and three from the massive vector fields (two transverse and one longitudinal polarisation states). Of course, a change of variables in the Lagrangian should not change the number of degrees of freedom of the system, so we must conclude that Eq. (5.35) contains non-physical field, which can be eliminated by so-called gauge fixing conditions.

If we write $\phi(x)$ in polar coordinates as in Eq. (5.12) and take a particular gauge transformation with $\alpha(x) = \frac{\eta(x)}{v}$, which is called unitary gauge transformation, we have

$$\phi(x) \to \phi'(x) = e^{\frac{-i\eta(x)}{v}}\phi(x) = \frac{1}{\sqrt{2}}(v + h(x)), \tag{5.36}$$

$$A_\mu(x) \to B_\mu(x) = A_\mu(x) - \frac{1}{ev}\partial_\mu \eta(x). \tag{5.37}$$

Under this unitary gauge transformation, the covariant derivative is transformed as

$$D_\mu \phi(x) \to D'_\mu \phi'(x) = (\partial_\mu - ieB_\mu)\frac{1}{\sqrt{2}}(v + h(x)) \tag{5.38}$$

and the field tensor transforms as

$$F_{\mu\nu}(A) = \partial_\nu A_\mu - \partial \mu A_\nu \to F_{\mu\nu}(B) = \partial_\nu B_\mu - \partial \mu B_\nu. \tag{5.39}$$

Then, the Lagrangian density in the unitary gauge becomes

$$\begin{aligned}
\mathscr{L} &= \frac{1}{2}|\partial_\mu h - ieB_\mu(v+h)|^2 - \frac{\mu^2}{2}(v+h)^2 - \frac{\lambda}{4}(v+h)^4 - \frac{1}{4}F_{\mu\nu}(B)F^{\mu\nu}(B) \\
&= \frac{1}{2}(\partial^\mu h)(\partial_\mu h) - \frac{1}{2}(2\mu^2)h^2 - \frac{1}{4}F_{\mu\nu}(B)F^{\mu\nu}(B) + \frac{1}{2}(ev^2)B_\mu B^\mu \\
&+ \frac{1}{2}e^2 B_\mu B^\mu h(h+2v) - \lambda v h^3 - \frac{\lambda}{4}h^4.
\end{aligned} \tag{5.40}$$

This Lagrangian describes a massive vector boson B_μ with mass $m_B = ev$ and a massive scalar boson h with mass $m_h = \sqrt{2\mu^2}$ which is called Higgs boson. We can see that the Goldstone boson $\eta(x)$ disappears and the massive vector boson $B_\mu(x)$ and Higgs boson $h(x)$ become physical. If we now count the number of degrees of freedom, we find that we begin with four degrees of freedom, as follows: two from the complex scalar field $\phi(x)$ and two from the massless vector boson A_μ. Then, after SSB in the unitary gauge, we also get four degrees of freedom, as follows: one from the real Higgs boson field $h(x)$ plus three from the massive vector boson $B_\mu(x)$. Thus, the number of degrees of freedom is conserved by the Higgs mechanism. We can illustratively say that the Goldstone boson was 'eaten up' by the vector boson B_μ, which in turn has become massive, as manifest from the new longitudinal polarisation component of it.

5.5.2 $SU(2)$ Gauge Symmetry

The Higgs mechanism can be extended to larger symmetries than $U(1)$. Let us turn our attention to a non-Abelian $SU(2)$ group. We consider a complex scalar field $\phi(x)$ belonging to the fundamental representation of the $SU(2)$ group

$$\phi = \begin{pmatrix} \phi_1 \\ \phi_2 \end{pmatrix}.$$

The $SU(2)$ local gauge invariant Lagrangian is given by

$$\mathscr{L} = (D_\mu\phi)^\dagger(D^\mu\phi) - \frac{1}{4}G^i_{\mu\nu}G^{i\mu\nu} - V(\phi^\dagger\phi), \tag{5.41}$$

where

$$D_\mu\phi = (\partial_\mu - ig\frac{\tau^i}{2}W^i_\mu)\phi \quad (i = 1, 2, 3), \tag{5.42}$$

$$G^i_{\mu\nu} = \partial_\mu W^i_\nu - \partial_\mu W^i_\mu + g\epsilon^{ijk}W^j_\mu W^k_\nu, \tag{5.43}$$

$$V(\phi^\dagger\phi) = \mu^2\phi^\dagger\phi + \lambda(\phi^\dagger\phi)^2. \tag{5.44}$$

For $\mu^2 < 0$, the vacuum is not unique and SSB can occur. We can write $\phi(x)$ in terms of four real fields $h(x)$ and $\eta^i(x)$ $(i = 1, 2, 3)$ as

$$\phi = \frac{1}{\sqrt{2}}e^{\frac{i\tau^i\eta^i}{2v}}\begin{pmatrix} 0 \\ v + h \end{pmatrix}. \tag{5.45}$$

We can also write our fields in the unitary gauge (as done in the $U(1)$ model) as

$$\phi \to \phi' = U\phi = \frac{1}{\sqrt{2}} \begin{pmatrix} 0 \\ v+h \end{pmatrix} \tag{5.46}$$

and the gauge fields, $W_\mu = W_\mu^i \frac{\tau^i}{2}$, in the unitary gauge become

$$W_\mu \to W'_\mu = UW_\mu U^{-1} - \frac{i}{g}(\partial_\mu U)U^{-1}, \tag{5.47}$$

where

$$U = e^{-i\tau^i \eta^i/2v}. \tag{5.48}$$

This leads to

$$D_\mu \phi \to D'_\mu \phi' = \left(\partial_\mu - ig\frac{\tau^i}{2} W'_\mu\right) \frac{1}{\sqrt{2}} \begin{pmatrix} 0 \\ v+h \end{pmatrix}, \tag{5.49}$$

$$G^i_{\mu\nu}(W)G^{i\mu\nu}(W) \to G^i_{\mu\nu}(W')G^{i\mu\nu}(W') = G^i_{\mu\nu}(W)G^{i\mu\nu}(W), \tag{5.50}$$

where

$$G^i_{\mu\nu} = \partial_\mu W^{i\prime}_\nu - \partial_\mu W^{i\prime}_\mu + g\epsilon^{ijk} W^{j\prime}_\mu W^{k\prime}_\nu. \tag{5.51}$$

We can see that the field strength tensor $G^i_{\mu\nu}$ is invariant under unitary gauge transformations. We can also see that, by writing our fields in the unitary gauge, the Goldstone bosons $\eta^i(x)$ $(i = 1, 2, 3)$ disappear and become the longitudinal components of the gauge fields $W^{'i}_\mu$ $(i = 1, 2, 3)$. The Lagrangian density in the unitary gauge becomes

$$\mathscr{L} = (D_\mu \phi)'^\dagger (D^\mu \phi)' - \frac{1}{4} G^i_{\mu\nu}(W')G^{i\mu\nu}(W') - \mu^2(\phi'^\dagger \phi') - \lambda(\phi'^\dagger \phi')^2. \tag{5.52}$$

From the covariant derivative, one obtains

$$\begin{aligned} [(D_\mu \phi)'^\dagger)]^a [(D^\mu \phi)']_a =& \frac{1}{2}\partial_\mu h \partial^\mu h + g^2 W'^i_\mu W'^{j\mu} \left(\frac{\tau^i}{2}\right)^a_b \left(\frac{\tau^j}{2}\right)^c_a \phi'^b \phi'_c \\ =& \frac{1}{2}\partial_\mu h \partial^\mu h + \frac{g^2}{8} W'^i_\mu W'^\mu (v+h)^2. \end{aligned} \tag{5.53}$$

Thus, the final form of the Lagrangian is given by

$$\mathscr{L} = \frac{1}{2}\partial_\mu h \partial^\mu h - \frac{1}{2}(2\mu^2)h^2 - \frac{1}{4}G^i_{\mu\nu}(W')G^{i\mu\nu}(W') + \frac{g^2 v^2}{8}W'^i_\mu W'^{i\mu}$$
$$+ \frac{g^2}{8}W'^i_\mu W'^\mu h(2v+h) - \lambda v h^3 - \frac{\lambda}{4}h^4. \tag{5.54}$$

This Lagrangian describes three massive vector bosons W'^i_μ $(i = 1, 2, 3)$ with mass $m_W = \frac{1}{2}gv$ and a massive scalar boson, *i.e.*, the Higgs boson $h(x)$ with mass $m_h = \sqrt{2\mu^2}$. We started with ten degrees of freedom: one from the real field $h(x)$, three Goldstone fields $\eta^i(x)$ $(i = 1, 2, 3)$ and six massless gauge fields W'^i_μ. We then end with the same number of degrees of freedom: one from the real scalar field $h(x)$ and nine from three massive vector bosons. We can therefore say that the three Goldstone bosons were eaten by the gauge fields W'^i_μ and have become their longitudinal polarisation components. In summary, the number of degrees of freedom is conserved by the Higgs mechanism also in a non-Abelian $SU(2)$ gauge theory.

CHAPTER 6

Glashow-Weinberg-Salam Model

After Glashow proposed the $SU(2)_L \times U(1)_Y$ model [27], Weinberg in 1967 [30] and independently Salam in 1968 [31] used the idea of SSB and Higgs mechanism [33–36] to generate masses for the weak gauge bosons, while keeping the photon massless as required. This led to a renormalisable theory of EW interactions, as proven by 't Hooft and Veltman from 1971 [37–39]. Therefore, the Glashow-Weinberg-Salam model was recognised as the SM of particle physics.

6.1 $SU(2)_L \times U(1)_Y$ SYMMETRY BREAKING

In Chapter 4 we have constructed the Lagrangian of weak interaction, which was based on the gauge invariance of $SU(2)_L \times U(1)_Y$. This theory is renormalisable; however, it is in disagreement with experimental data. In this model, all fermions and gauge bosons ($W^{\pm}, Z$ and γ) are massless (in order to retain the gauge invariance). In order to construct a correct renormalisable theory of weak interactions for physical particles, we must implement the Higgs mechanism to break the $SU(2)_L \times U(1)_Y$ symmetry spontaneously, so as to generate masses for SM fermions and weak gauge bosons. As $U(1)_{\text{EM}}$ is an exact symmetry of our world (indeed, with a massless photon), we are looking at the following symmetry breaking pattern:

$$SU(2)_L \times U(1)_Y \to U(1)_{\text{EM}}. \tag{6.1}$$

DOI: 10.1201/9780429443015-6

To this end, we introduce a complex scalar field $\phi(x)$ that interacts with the EW gauge bosons, *i.e.*, the $W^\pm$, Z and γ states. As a result of the Higgs mechanism and the violation of gauge invariance, as explained in the previous chapter, three vector particles ($W^\pm$ and Z) will acquire masses. This means that the Higgs field must form at least a doublet. Thus, we suppose that the Higgs field is given by

$$\phi = \begin{pmatrix} \phi^+ \\ \phi^0 \end{pmatrix}, \tag{6.2}$$

where ϕ^+ and ϕ^0 represent charged and neutral complex scalar fields, respectively. In this case, one can easily show that the weak hypercharge is given by $Y_\phi = +1$ and the doublet $\phi(x)$ has a well-defined transformation under $SU(2)_L$ and $U(1)_Y$, as follows. Under $SU(2)_L$ one has

$$\begin{aligned} \phi(x) \to \phi'(x) &= e^{-\frac{ig}{2}\tau_j \alpha_j(x)} \phi(x), \\ \phi^\dagger(x) \to \phi'^\dagger(x) &= \phi^\dagger(x) e^{\frac{ig}{2}\tau_j \alpha_j(x)}, \end{aligned} \tag{6.3}$$

while under $U(1)_Y$ one has

$$\begin{aligned} \phi(x) \to \phi'(x) &= e^{-i\frac{g'}{2}\beta(x)} \phi(x), \\ \phi^\dagger(x) \to \phi'^\dagger(x) &= \phi^\dagger(x) e^{i\frac{g'}{2}\beta(x)}, \end{aligned} \tag{6.4}$$

where g and g' are the gauge couplings associated with $SU(2)_L$ and $U(1)_Y$, respectively. Then, a local $SU(2)_L \times U(1)_Y$ gauge invariant Lagrangian of this complex scalar field $\phi(x)$ is

$$\mathscr{L}_{\text{scalar}} = (D_\mu \phi)^\dagger)(D^\mu \phi) - \mu^2 |\phi|^2 - \lambda |\phi|^4, \tag{6.5}$$

where

$$D_\mu \phi = \left(\partial_\mu - ig\frac{\tau^i}{2} W_\mu^i - \frac{i}{2} g' B_\mu \right) \phi. \tag{6.6}$$

We again take $\lambda > 0$ to ensure that our system will be bounded from below (*i.e.*, with a stable vacuum state). As usual, higher order terms in $\phi(x)$ in the above Lagrangian are not allowed by renormalisability conditions. For $\mu^2 < 0$, a non-vanishing vacuum can be obtained and SSB occurs. This minimum of $\phi(x)$ is given by

$$(\phi^\dagger \phi)_0 = \frac{v^2}{2}, \quad \text{where} \quad v = \sqrt{\frac{\mu^2}{\lambda}}. \tag{6.7}$$

Let us choose

$$\phi_0 = \begin{pmatrix} 0 \\ \frac{v}{\sqrt{2}} \end{pmatrix}. \tag{6.8}$$

It is clear that the neutral component of $\phi(x)$ only is getting a non-zero VEV to ensure that $U(1)_{\rm EM}$ remains unbroken. One can show explicitly that the generators of $SU(2)_L$ and $U(1)_Y$ do not annihilate the vacuum ϕ_0, since

$$\begin{aligned} \frac{1}{2}\tau^1\phi_0 &= \frac{1}{2}\begin{pmatrix} \frac{v}{\sqrt{2}} \\ 0 \end{pmatrix}, \\ \frac{1}{2}\tau^2\phi_0 &= \frac{1}{2}\begin{pmatrix} \frac{-iv}{\sqrt{2}} \\ 0 \end{pmatrix}, \\ \frac{1}{2}\tau^3\phi_0 &= -\frac{1}{2}\phi_0, \\ Y\phi_0 &= \phi_0. \end{aligned} \tag{6.9}$$

Thus, the $SU(2)_L \times U(1)_Y$ symmetry is spontaneously broken by ϕ_0 while the electric charge operator Q annihilates the vacuum,

$$Q\phi_0 = \frac{1}{2}(\tau^3 + Y)\phi_0 = 0, \tag{6.10}$$

so that $U(1)_{\rm EM}$ remains an exact symmetry.

As we did in the previous section, we can parametrise a complex scalar field $\phi(x)$ in terms of four real fields, $\eta^i(x)$ $(i = 1, 2, 3)$ and $h(x)$, as follows:

$$\phi = e^{\frac{i\tau^i\eta^i}{2v}} \begin{pmatrix} 0 \\ \frac{v+h}{\sqrt{2}} \end{pmatrix}. \tag{6.11}$$

By writing our Lagrangian in the unitary gauge, we will see that the Goldstone bosons $\eta^i(x)$ $(i = 1, 2, 3)$ will disappear and will become th longitudinal components of $W^\pm$ and Z. As we did in the previous section, we apply the $SU(2)_L$ unitary gauge transformation

$$U(\eta) = e^{\frac{-i\tau^i\eta^i}{2v}}. \tag{6.12}$$

Our fields in the unitary gauge then become

$$\phi' = U(\eta)\phi = \begin{pmatrix} 0 \\ \frac{v+h}{\sqrt{2}} \end{pmatrix}, \tag{6.13}$$

$$f'_L = U(\eta) f_L, \tag{6.14}$$

$$f'_R = f_R, \tag{6.15}$$

$$W'_\mu = U(\eta) W_\mu U^\dagger(\eta) - \frac{i}{g}(\partial_\mu U(\eta)) U^\dagger(\eta), \tag{6.16}$$

where $W_\mu = \vec{W}_\mu \cdot \frac{\vec{\tau}}{2} = W^i_\mu \frac{\tau^i}{2}$.

6.2 EW GAUGE BOSON MASSES

As shown previously, the gauge bosons mass terms come from the covariant derivative of the scalar field after SSB. This can be seen in our case as follows:

$$(D^\mu \phi)' = \partial^\mu \phi' - \frac{i}{2}(g\tau^i W'^i + g' B^\mu)\phi'. \tag{6.17}$$

Here, we use prime symbols for fields in gauge eigenstates so that the usual unprimed fields can be defined in mass eigenstates. Therefore, one finds

$$(D_\mu \phi)'^\dagger (D^\mu \phi)' = \frac{1}{2}\partial_\mu h \partial^\mu h + \frac{1}{8}\phi'^\dagger (g W'_\mu \tau^i + g' B_\mu)(g\tau^j W'^{j\mu} + g' B^\mu)\phi'. \tag{6.18}$$

We then use the identity $\tau^i \tau^j = \delta^{ij} + i\epsilon^{ijk}\tau^k$ to obtain

$$\begin{aligned}
(D_\mu \phi)'^\dagger (D^\mu \phi)' &= \frac{1}{2}\partial_\mu h \partial^\mu h + \frac{v^2}{8}(g^2 W'^i_\mu W'^{i\mu} + g'^2 B'_\mu B'^\mu - 2gg' W'^3_\mu B'^\mu) \\
&+ \frac{1}{8}(2vh + h^2)(g^2 W'^i_\mu W'^{i\mu} + g'^2 B'_\mu B'^\mu - 2gg' W'^3_\mu B'^\mu) \\
&= \frac{1}{2}\partial_\mu h \partial^\mu h + \frac{v^2}{8}[g^2(W'^1_\mu W'^{1\mu} + W'^2_\mu W'^{2\mu}) + g^2 W'^3_\mu W'^{3\mu} \\
&- 2gg' W'^3_\mu B'^\mu + g'^2 B'_\mu B'^\mu] \\
&+ \frac{1}{8}(2vh + h^2)[g^2(W'^1_\mu W'^{1\mu} + W'^2_\mu W'^{2\mu}) + g^2 W'^3_\mu W'^{3\mu} \\
&- 2gg' W'^3_\mu B'^\mu + g'^2 B'_\mu B'^\mu].
\end{aligned} \tag{6.19}$$

We then collect mass terms, which are quadratic in the fields

$$\mathscr{L}_{\text{mass}} = \frac{v^2}{8}[g^2 W'^1_\mu W'^{1\mu} + g^2 W'^2_\mu W'^{2\mu} + (gW'^3_\mu - g' B'_\mu)^2]. \tag{6.20}$$

Upon defining two charged complex gauge fields, $W^+_\mu(x)$ and $W^-_\mu(x)$, as

$$W^\pm_\mu = \frac{1}{\sqrt{2}}(W^1_\mu \mp iW^2_\mu), \tag{6.21}$$

then $\mathscr{L}_{\text{mass}}$ becomes

$$\begin{aligned}\mathscr{L}_{\text{mass}} &= \frac{1}{4}g^2 v^2 W^+_\mu W^{-\mu} + \frac{v^2}{8}(gW'^3_\mu - g' B'_\mu)^2 \\ &= \frac{1}{4}g^2 v^2 W^+_\mu W^{-\mu} + \frac{v^2}{8}\begin{pmatrix} W'^3_\mu & B'_\mu \end{pmatrix}\begin{pmatrix} g^2 & -2gg' \\ -2gg' & g'^2 \end{pmatrix}\begin{pmatrix} W'^\mu \\ B'^\mu \end{pmatrix}.\end{aligned} \tag{6.22}$$

The first term in $\mathscr{L}_{\text{mass}}$ describes the mass contribution common to the two charged gauge bosons $W^\pm$ while the mass matrix which describes the masses of the neutral gauge fields W'^3_μ and B'_μ is non-diagonal, as explicitly described by the second term in Eq. (6.22). However, we can diagonalise this mass matrix by making a unitary transformation of W'^3_μ and B'_μ, as follows:

$$\begin{pmatrix} W'^3_\mu \\ B_\mu \end{pmatrix} = \begin{pmatrix} \cos\theta_W & \sin\theta_W \\ -\sin\theta_W & \cos\theta_W \end{pmatrix}\begin{pmatrix} Z_\mu \\ A_\mu \end{pmatrix}, \tag{6.23}$$

where θ_W is the already introduced Weinberg (or weak) mixing angle, so that the two fields Z_μ and A_μ, upon quantisation, give rise to the neutral EW gauge boson, indeed the massive Z boson and the massless photon.

We now use the relation which links the gauge couplings g and g' and Weinberg angle θ_W,

$$g\sin\theta_W = g'\cos\theta_W, \tag{6.24}$$

or

$$\sin\theta_W = \frac{g'}{\sqrt{g^2 + g'^2}}, \quad \cos\theta_W = \frac{g}{\sqrt{g^2 + g'^2}}, \tag{6.25}$$

so, finally, $\mathscr{L}_{\text{mass}}$ becomes

$$\begin{aligned}\mathscr{L}_{\text{mass}} &= \frac{1}{4}g^2v^2W_\mu^+W^{-\mu} + \frac{v^2}{8}\begin{pmatrix} Z_\mu & A_\mu \end{pmatrix}\begin{pmatrix} g^2+g'^2 & 0 \\ 0 & 0 \end{pmatrix}\begin{pmatrix} Z^\mu \\ A^\mu \end{pmatrix} \\ &= \frac{1}{4}g^2v^2W_\mu^+W^{-\mu} + \frac{v^2}{8}(g^2+g'^2)Z_\mu Z^\mu. \end{aligned} \tag{6.26}$$

We therefore have three massive spin-1 particles, $W^\pm$, Z and a photon γ still massless, indeed as in real world, with

$$M_W = \frac{1}{2}gv, \tag{6.27}$$

$$M_Z = \frac{1}{2}v\sqrt{g^2+g'^2} = \frac{M_W}{\cos\theta_W}, \tag{6.28}$$

$$M_A = 0. \tag{6.29}$$

The values of M_W and M_Z depend on θ_W, the value of $\sin^2\theta_W$ is obtained from neutrino scattering experiments to be around 0.23, so that the values of $M_W \simeq 80$ GeV and $M_Z \simeq 90$ GeV first measured by the Super Proton-Antiproton Synchrotron (S$p\bar{p}$S) at CERN are perfectly self-consistent.

6.3 HIGGS BOSON MASS

The potential energy density in the Lagrangian of the complex scalar field $\phi(x)$ becomes

$$\begin{aligned} V(\phi'^\dagger\phi') &= \frac{-\mu^2}{2}(v+h)^2 + \frac{\lambda}{4}(v+h)^4 \\ &= \frac{1}{2}(2\lambda v^2)h^2 + \lambda v h^3 + \frac{\lambda}{4}h^4 - \frac{\mu^2v^2}{4}. \end{aligned} \tag{6.30}$$

From Eq. (6.30), the mass of the Higgs boson field $h(x)$ is

$$M_h = \sqrt{2\lambda v^2}. \tag{6.31}$$

The scalar Lagrangian becomes

$$\begin{aligned}\mathscr{L}_{\text{scalar}} &= \frac{1}{2}\partial_\mu h \partial^\mu h - \frac{1}{2}M_h^2 h^2 - \lambda v h^3 - \frac{\lambda}{4}h^4 \\ &+ \frac{g^2}{8}(2vh + h^2)\left(\frac{1}{\cos^2\theta_W} Z_\mu Z^\mu + 2W_\mu^+ W^{-\mu}\right) \\ &+ M_W^2 W_\mu^+ W^{-\mu} + \frac{1}{2}M_Z^2 Z_\mu Z^\mu + \frac{\mu^2 v^2}{4}. \end{aligned} \tag{6.32}$$

The second line in the expression for $\mathscr{L}_{\text{scalar}}$ describes the interactions between the massive gauge and Higgs bosons. The Higgs boson of the SM was finally discovered at the LHC at CERN in 2012.

CHAPTER 7

Strong Interactions and QCD

The importance of QCD as the correct theory of strong interactions cannot be understated enough. In fact, it is a QFT with unique characteristics. On the one hand, it features 'asymptotic freedom'. On the other hand, it features 'infrared slavery' (or 'confinement'). It seems an oxymoron, possibly a contradiction in itself. Yet, it makes for one of the most fascinating presentations of Nature.

When probed at very short wavelengths, QCD is essentially a theory of free 'partons' (called quarks and gluons) which only scatter off one another through relatively small QFT effects, that can then be systematically calculated. But at longer wavelengths, of order the size of the proton ~ 1 fm $= 10^{-15}$ m, we only detect bound states of these partons (called hadrons) with string-like potentials building up if we try to separate their partonic constituents, see Fig. 7.1. That is, as, *e.g.*, quarks and antiquarks separate from each other the force between them grows, so does the energy stored in the system until it will become energetically favourable to create a quark-antiquark pair in the middle. The hadron has then broken into two which can then be separated at no extra energy cost. In short, never is a quark freed. This simple picture explains confinement. In fact, by the same argument, we can also understand jet formation. The quark-antiquark pairs are created close to each other and behave weakly. As they move apart the coupling grows and the energy of that interaction then creates more quark-antiquark pairs. We thus see hadronic jets not the original quarks.

DOI. 10.1201/9780429443015-7

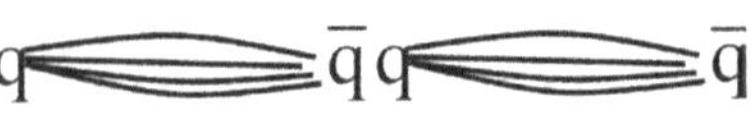

Figure 7.1 Pictorial representation of the confinement mechanism, where generic field lines are also sketched.

Due to our inability to perform calculations in strongly coupled field theories, QCD is therefore still only partially solved. Nonetheless, all its features, across all energy scales, are encoded in a single one-line formula, *i.e.*, the QCD Lagrangian.

Starting from the latter, in this chapter, we will discuss the basics of $SU(3)_C$, colour factors and the running of α_s. For many parts of this chapter, we are greatly indebted to the clear formulation of Ref. [40].

7.1 LAGRANGIAN OF QCD

The underlying structure of QCD is a $SU(3)_C$ gauge symmetry, with $N_C = 3$ being the number of colours[1], *i.e.*, the special unitary group in 3 (complex) dimensions which elements are unitary matrices with dimensions 3×3 and determinant equal to 1. Since there are 9 linearly independent unitary complex matrices[2], one of which has determinant -1, there are a total of 8 independent directions in this matrix space, corresponding to 8 different generators (recall that in the case of QED there is only 1). In QCD, one usually represents such a gauge group exploiting the so-called *fundamental* representation, wherein the generators of $SU(3)_C$ appear as a set of 8 traceless Hermitian matrices, to which we return below.

We shall refer to indices enumerating rows and columns of these matrices $(1, ..., 3)$ as *fundamental* indices and we adopt the letters i, j, k, ... to denote them. We refer to indices enumerating the generators $(1, ..., 8)$ as *adjoint* indices[3], and we use the first letters of the alphabet (a, b, c, ...) to denote them.

[1] Which we characterise as R, and B for Red, Green and Blue, respectively.

[2] A $N \times N$ complex matrix has $2N^2$ degrees of freedom with unitarity providing N^2 constraints.

[3] The dimension of the *adjoint* representation is equal to the number of generators,

These matrices can operate both on each other and on a set of 3-vectors, the latter being representations of (anti)quarks in colour space, which are *triplets* under $SU(3)_C$. These matrices are representing gluons in colour space; hence, there are 8 different ones (*i.e.*, they are *octets* under $SU(3)_C$).

The aforementioned Lagrangian (density) of QCD is written as

$$\mathcal{L} = \bar{\psi}_q^i (i\gamma^\mu)(D_\mu)_{ij}\psi_q^j - m_q \bar{\psi}_q^i \psi_{qi} - \frac{1}{4} F_{\mu\nu}^a F^{a\mu\nu}, \tag{7.1}$$

where $\psi_{q\,i}$ denotes a quark field with (fundamental) colour index $i = R, G, B$, $\psi_q = (\psi_{qR}, \psi_{qG}, \psi_{qB})^T$, γ^μ is a Dirac matrix that expresses the vector nature of the strong interaction, with μ being a Lorentz vector index, m_q is a quark mass, $F_{\mu\nu}^a$ is the field strength tensor for a gluon[4] with (adjoint) colour index a (*i.e.*, $a = 1, ..., 8$) and D_μ is the covariant derivative in QCD,

$$(D_\mu)_{ij} = \delta_{ij}\partial_\mu - ig_s t_{ij}^a A_\mu^a, \tag{7.2}$$

with g_s the strong coupling (related to α_s by $g_s^2 = 4\pi\alpha_s$), A_μ^a is the gluon field with colour index a and the t_{ij}^a matrices are proportional to the Hermitian and traceless Gell-Mann ones λ^a of $SU(3)_C$, which we have already introduced. (These generators are just the QCD version of the Pauli matrices in $SU(2)_L$.) Conventionally, the latter are related as follows:

$$t_{ij}^a = \frac{1}{2}\lambda_{ij}^a. \tag{7.3}$$

This choice in turn determines the normalisation of the QCD coupling g_s, via Eq. (7.2), and fixes the values of the $SU(3)_C$ Casimir factors (Casimirs for short) and structure constants, to which we will return below.

An example of a possible colour flow for a quark-gluon interaction in colour space is given in Fig. 7.2, wherein, in computing physics observables, we have to sum over all colour indices, so this sketch merely gives a pictorial representation of one particular (non-zero) term in such a colour sum.

$N^2 - 1 = 8$ (for $SU(3)_C$), while the dimension of the fundamental representation is the degree of the group, $N = 3$ (for $SU(3)_C$).

[4]The definition of the gluon field strength tensor will be given below in Eq. (7.8).

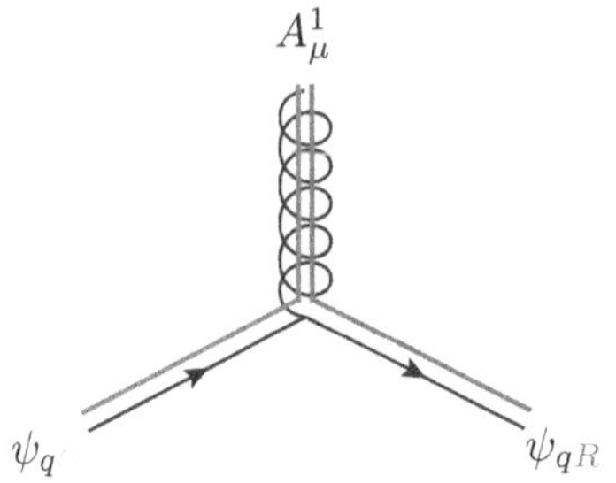

A^1_μ

$$\propto \quad -\tfrac{i}{2}g_s \quad \bar{\psi}_{qR} \quad \lambda^1 \quad \psi_{qG}$$

$$= \quad -\tfrac{i}{2}g_s \quad \begin{pmatrix} 1 & 0 & 0 \end{pmatrix} \begin{pmatrix} 0 & 1 & 0 \\ 1 & 0 & 0 \\ 0 & 0 & 0 \end{pmatrix} \begin{pmatrix} 0 \\ 1 \\ 0 \end{pmatrix}$$

ψ_{qG} ψ_{qR}

Figure 7.2 A qqg vertex in QCD, before summing/averaging over colours: a gluon in a state represented by λ^1 interacts with quarks in the states ψ_{qR} and ψ_{qG} (from Ref. [40]).

7.2 COLOUR FACTORS

As one cannot extract colours from physics observables, we average over all possible incoming ones and sum over all possible outgoing ones, so that QCD scattering amplitudes (squared) always contain sums over quark fields contracted with Gell-Mann matrices. These contractions then produce traces which yield the *colour factors* that are associated with each QCD process and which basically count the number of "paths (or trajectories) in colour space" that the process considered can take.

A useful example of a colour factor is obtained from the process $e^+e^- \to \gamma^*, Z \to q\bar{q}$, see Fig. 7.3 (left), a process to which we will come back later on at length. The corresponding vertex contains a simple Kronecker δ_{ij} in colour space: *i.e.*, the outgoing quark and antiquark must have identical (anti)colours. Squaring the corresponding ME and summing over final state colours yields a colour factor of

$$e^+e^- \to \gamma^*, Z \to q\bar{q}: \qquad \sum_{\text{colours}} |M|^2 \propto \delta_{ij}\delta_{ji} = \text{Tr}\{\delta\} = 3, \tag{7.4}$$

since i and j are (anti)quark (*i.e.*, 3-dimensional fundamental) indices. (Here, there is no need to average over colours in the initial state, naturally.) This factor corresponds directly to the N_C different paths through colour space that the process can take, *i.e.*, the produced (anti)quark can be R, G or B in (anti)colour.

A correlated example is given by the s-channel $q\bar{q} \to \gamma^*, Z \to e^+e^-$ process, which is just a simple time reversal of the previous one, see Fig. 7.3

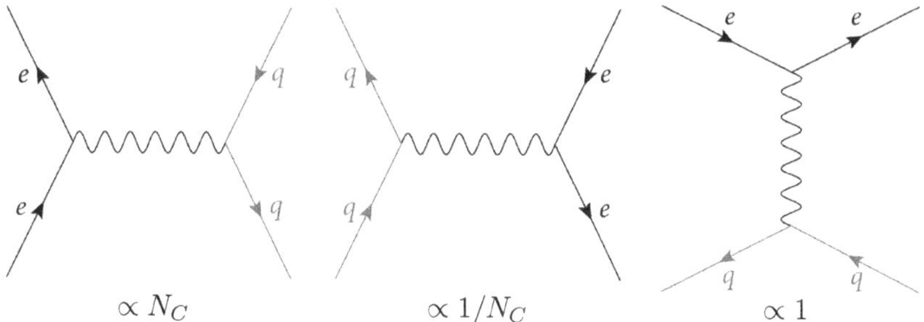

Figure 7.3 The three crossings of the basic interaction between an electron(positron) current (in Black) and an (anti)quark one (in Red) via an intermediate γ or Z boson, with corresponding colour factors (from Ref. [40]).

(centre). By crossing symmetry, the squared ME, including the colour factor, is exactly the same as before but, since the quarks are here incoming, we must *average* rather than sum over their colours, leading to

$$q\bar{q} \to \gamma^*, Z \to e^+e^- : \qquad \frac{1}{9}\sum_{\text{colours}} |M|^2 \propto \frac{1}{9}\delta_{ij}\delta_{ji} = \frac{1}{9}\text{Tr}\{\delta\} = \frac{1}{3}, \qquad (7.5)$$

where the colour factor now represents a *suppression*, which can be interpreted as due to the fact that only quark-antiquark pairs of matching (anti)colours are able to annihilate and produce a (colourless) γ^*, Z intermediate state. The chance that a quark and an antiquark extracted at random from colliding hadrons have matching colours is, in fact, $1/N_C$.

Similarly, the DIS $e^\pm q \to e^\pm q$ (or, similarly, $e^\pm \bar{q} \to e^\pm \bar{q}$) process via t-channel γ^*, Z^* exchange constitutes yet another crossing of the same basic process, see Fig. 7.3 (right). The colour factor in this case is 1, exemplifying the fact that, no matter the colour of the (anti)quark, this interaction will always take place.

To illustrate what happens when we look at processes in which gluons are involved, so that we have to insert (and sum over) quark-gluon vertices, such as the one depicted in Fig. 7.2, we take the $Z \to q\bar{q}g$ decay. (This is similar to the $e^+e^- \to \gamma^*, Z \to q\bar{q}$ process already considered, though we take now only the Z (on-shell) and factorise out its (radiative) decay, see later on.) The colour factor for this process can be computed through the following colour

Trace Relation	Indices	Occurs in Diagram Squared
$\mathrm{Tr}\{t^a t^b\} = T_R\,\delta^{ab}$	$a, b \in [1, \ldots, 8]$	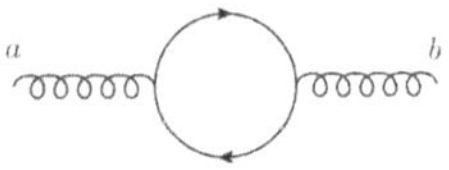
$\sum_a t^a_{ij} t^a_{jk} = C_F\,\delta_{ik}$	$a \in [1, \ldots, 8]$ $i, j, k \in [1, \ldots, 3]$	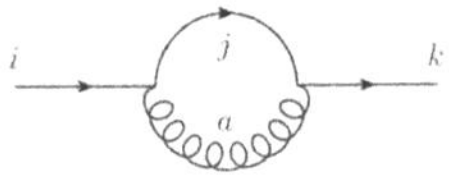
$\sum_{c,d} f^{acd} f^{bcd} = C_A\,\delta^{ab}$	$a, b, c, d \in [1, \ldots, 8]$	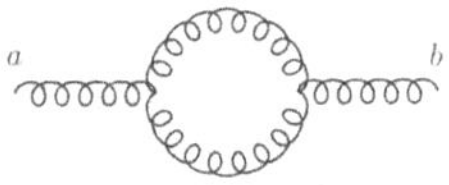
$t^a_{ij} t^a_{kl} = T_R \left(\delta_{jk}\delta_{il} - \frac{1}{N_C}\delta_{ij}\delta_{kl} \right)$	$i, j, k, l \in [1, \ldots, 3]$	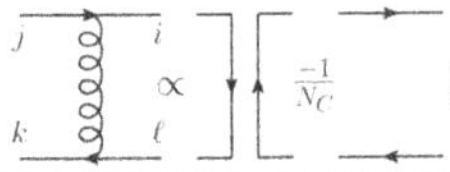 (Fierz)

Table 7.1 Trace relations for t^a matrices (from Ref. [40]). More such relations can be found in, *e.g.*, Refs. [41, 42].

diagram (squared) with explicit colour indices in each vertex:

$Z \to q\bar{q}g$:

$$\sum_{\text{colours}} |M|^2 \propto \delta_{ij} t^a_{jk} t^a_{kl} \delta_{li} = \mathrm{Tr}\{t^a t^a\} = \frac{1}{2}\mathrm{Tr}\{\delta\} = 4, \qquad (7.6)$$

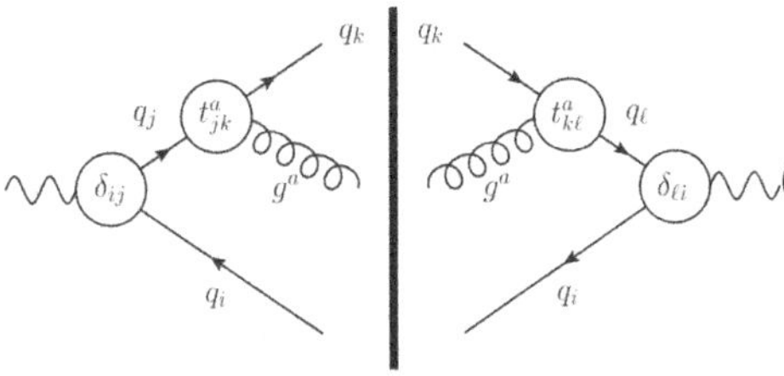

where $\mathrm{Tr}\{\delta\} = 8$ since the trace runs over the 8-dimensional adjoint indices. If we want to count the usual paths through colour space, we should leave out the factor $\frac{1}{2}$ which comes from the normalisation convention for the t^a matrices, Eq. (7.3), so that this process can take 8 different such trajectories, indeed, one per each gluon state.

The repetitive task of taking traces over t^a matrices can be greatly alleviated by using the relations given in Tab. 7.1. In the standard normalisation convention for the $SU(3)_C$ generators, Eq. (7.3), the Casimirs of $SU(3)_C$ appearing in Tab. 7.1 are[5]

$$T_R = \frac{1}{2}, \qquad C_F = \frac{4}{3}, \qquad C_A = N_C = 3. \qquad (7.7)$$

[5] In literature, sometimes, T_F is used in place of T_R.

(Notice that they essentially measure the strength of the interactions appearing in amplitudes squared, as sketched in the last column of Tab. 7.1.)

The first two Casimirs are solely expressed in terms of the t^a matrices. In contrast, the third Casimir involves factors of f^{abc}. These are the so-called *structure constants* of QCD and they enter via the non-Abelian term in the gluon field strength tensor appearing in Eq. (7.1),

$$F^a_{\mu\nu} = \underbrace{\partial_\mu A^a_\nu - \partial_\nu A^a_\mu}_{\text{Abelian}} + \underbrace{g_s f^{abc} A^b_\mu A^c_\nu}_{\text{non-Abelian}} . \tag{7.8}$$

Such structure constants of $SU(3)_C$ are given as follows:

$$f_{123} = 1, \tag{7.9}$$

$$f_{147} = f_{246} = f_{257} = f_{345} = \frac{1}{2}, \tag{7.10}$$

$$f_{156} = f_{367} = -\frac{1}{2}, \tag{7.11}$$

$$f_{458} = f_{678} = \frac{\sqrt{3}}{2}, \tag{7.12}$$

which are antisymmetric in all indices, while for all others one has $f_{abc} = 0$.

They define the *adjoint* representation of $SU(3)_C$ and are related to the fundamental representation generators via the commutator relations

$$t^a t^b - t^b t^a = [t^a, t^b] = i f^{abc} t_c, \tag{7.13}$$

or, equivalently,

$$i f^{abc} = 2\mathrm{Tr}\{t^c [t^a, t^b]\}. \tag{7.14}$$

Therefore, it is just a matter of choice whether to express the colour space in terms of the t^a matrices or via the f^{abc} structure constants, they are interchangeable. Expanding the $F_{\mu\nu}F^{\mu\nu}$ term of the Lagrangian using Eq. (7.8), we see that there are both 3- and 4-gluon vertices that involve f^{abc}, the latter of which has two powers of f^{abc} and two powers of the coupling g_s. Finally, the last line of Tab. 7.1 is not really a trace relation but instead a useful so-called *Fierz transformation*, which expresses products of t^a matrices in terms of Kronecker δ functions and recurs often in QCD calculations in colour space.

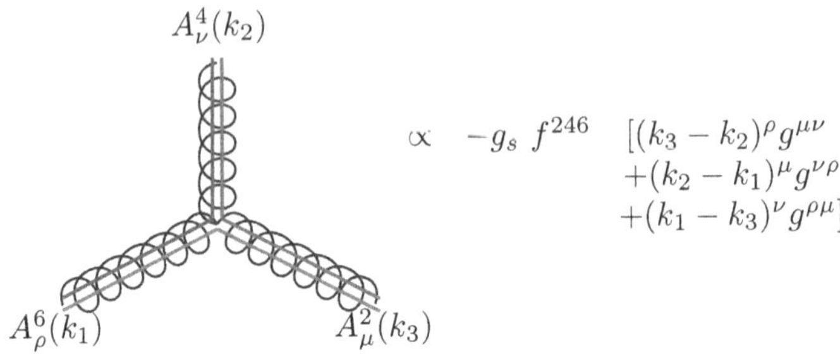

Figure 7.4 A ggg vertex in QCD before summing/averaging over colours, wherein the interaction between gluons in the states λ^2 $(R, \ )$, λ^4 (B, R) and λ^6 $(\ , B)$ is represented by the structure constant f^{246} (from Ref. [40]).

A gluon self-interaction triple vertex is illustrated in Fig. 7.4[6], to be compared with the quark-gluon one in Fig. 7.2. We remind the reader that gauge boson self-interactions are a hallmark of non-Abelian gauge theories and that their presence leads to some of the main differences between QED and QCD, as we shall discuss in the next chapter.

7.3 STRONG COUPLING

Such a gluon self-interaction is responsible for the peculiar running of the QCD coupling constant α_s. The latter is logarithmic with energy and is governed by the equation

$$Q^2 \frac{\partial \alpha_s}{\partial Q^2} = \frac{\partial \alpha_s}{\partial \ln Q^2} = \beta(\alpha_s), \tag{7.15}$$

where the function driving the energy dependence, the so-called *β-function*, is defined as

$$\beta(\alpha_s) = -\alpha_s^2 (b_0 + b_1 \alpha_s + b_2 \alpha_s^2 + \ldots), \tag{7.16}$$

with LO (one-loop) and NLO (two-loop) coefficients

$$b_0 = \frac{11 C_A - 4 T_R n_f}{12\pi}, \tag{7.17}$$

$$b_1 = \frac{17 C_A^2 - 10 T_R C_A n_f - 6 T_R C_F n_f}{24\pi^2} = \frac{153 - 19 n_f}{24\pi^2}. \tag{7.18}$$

[6]We do not dwell on the quadruple (or quartic) gluon vertex here, as it can be written as a superposition of three triple ones.

In the b_0 coefficient, the first term is due to gluon loops while the second is due to quark ones. Similarly, the first term of the b_1 coefficient arises from double gluon loops while the second and third represent mixed quark-gluon ones. (At higher loop orders, the b_i coefficients depend explicitly on the renormalisation scheme that is used, so we do not present these here.)

Numerically, the strong coupling is usually specified by giving its value at a specific reference scale, *e.g.*, $Q^2 = M_Z^2$, from which one can obtain its value at any other scale by solving Eq. (7.15),

$$\alpha_s(Q^2) = \alpha_s(M_Z^2)\frac{1}{1 + b_0\alpha_s(M_Z^2)\ln\frac{Q^2}{M_Z^2} + \mathcal{O}(\alpha_s^2)}, \tag{7.19}$$

with relations including the $\mathcal{O}(\alpha_s^2)$ terms available, *e.g.*, in Ref. [41]. Relations between scales not involving M_Z^2 can obviously be obtained by just replacing M_Z^2 by some other scale Q'^2 everywhere in Eq. (7.19).

The appearance of the number of flavours, n_f, in b_0 implies that the slope of the running depends on the number of contributing flavours. Since full QCD is best approximated by $n_f = 3$ below the charm threshold, by $n_f = 4$ and 5 from there to the b- and t-quark thresholds, respectively, and then by $n_f = 6$ at scales higher than m_t, it is therefore important to be aware that the running changes slope across quark flavour thresholds.

The negative overall sign of Eq. (7.16), combined with the fact that $b_0 > 0$ (for $n_f \leq 16$), leads to the famous result that the QCD coupling effectively *decreases* with energy, called asymptotic freedom, for the discovery of which the Nobel prize in physics was awarded to Gross, Politzer and Wilczek in 2004 [43, 44]. Among the consequences of asymptotic freedom is that perturbation theory becomes better behaved at higher absolute energies, due to the effectively decreasing coupling. Our current understanding of the running of the QCD coupling is summarised by the plot in Fig. 7.5, taken from the review in Ref. [45].

As a final remark on asymptotic freedom, note that the decreasing value of the strong coupling with energy must eventually cause it to become comparable to the EM and weak ones, at some energy scale. Beyond that point, which may lie at energies of order $10^{15} - 10^{17}$ GeV, we do not know what the further evolution of the combined theory will actually look like, or whether it will continue to exhibit asymptotic freedom.

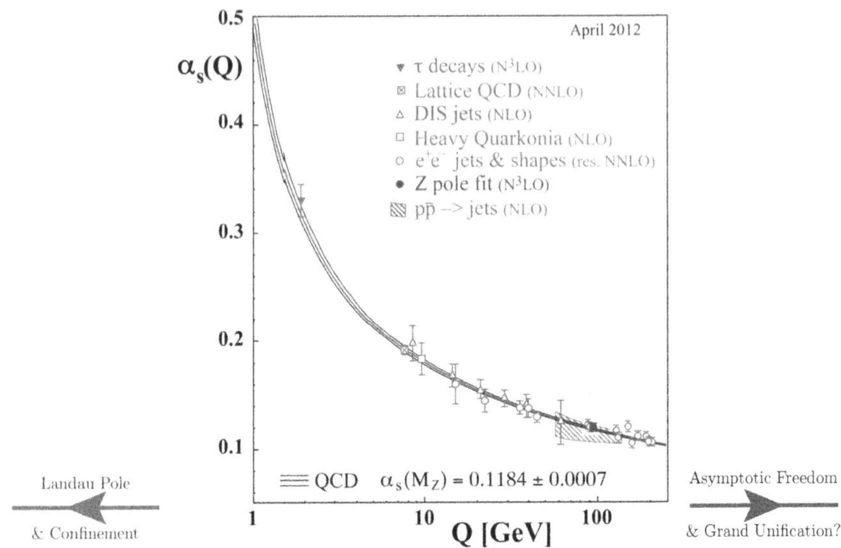

Figure 7.5 The running of α_s in a theoretical calculation (band) and in physical processes at different energies, from [45]. The little kinks at $Q = 2m_c$ and $Q = 2m_b$ are caused by discontinuities in the running across the flavour thresholds (from Ref. [40]).

Now consider what happens when we run the coupling in the other direction, towards smaller energies. Taken at face value, the numerical value of the coupling diverges rapidly at scales below 1 GeV, the so-called *hadronisation scale*, as illustrated by the curves disappearing off the left-hand edge of the plot in Fig. 7.5. To make this divergence explicit, one can rewrite Eq. (7.19) in the following form:

$$\alpha_s(Q^2) = \frac{1}{b_0 \ln \frac{Q^2}{\Lambda^2_{\mathrm{QCD}}}}, \tag{7.20}$$

where

$$\Lambda_{\mathrm{QCD}} \sim 200\,\mathrm{MeV} \tag{7.21}$$

specifies the energy scale at which the perturbative coupling would nominally become infinite, called the Landau pole. (Note, however, that this only parametrises the purely *perturbative* result, which is not reliable for a strong coupling, so Eq. (7.20) should not be taken to imply that the physical behaviour of full QCD should exhibit a divergence for $Q \to \Lambda_{\mathrm{QCD}}$.)

Finally, one should be aware that there is a multitude of different ways of defining both Λ_{QCD} and $\alpha_s(M_Z)$. At the very least, the numerical value one obtains depends both on the renormalisation scheme used (see later on) and on the perturbative order of the calculations used to extract them. As a rule of thumb, fits to experimental data typically yield smaller values for $\alpha_s(M_Z)$ the higher the order of the calculation used to extract it (see, *e.g.*, Refs. [45, 46]), with $\alpha_s(M_Z)|_{\mathrm{LO}} \gtrsim \alpha_s(M_Z)|_{\mathrm{NLO}} \gtrsim \alpha_s(M_Z)|_{\mathrm{NNLO}}$. Further, since the number of flavours changes the slope of the running, the location of the Landau pole for fixed $\alpha_s(M_Z)$ depends explicitly on the number of flavours used in the running. Thus each value of n_f is associated with its own value of Λ_{QCD}, with the following matching relations across thresholds guaranteeing continuity of the QCD coupling at one loop:

$$n_f = 5 \leftrightarrow 6: \qquad \Lambda_6 = \Lambda_5 \left(\frac{\Lambda_5}{m_t}\right)^{\frac{2}{21}}, \qquad \Lambda_5 = \Lambda_6 \left(\frac{m_t}{\Lambda_6}\right)^{\frac{2}{23}}, \tag{7.22}$$

$$n_f = 4 \leftrightarrow 5: \qquad \Lambda_5 = \Lambda_4 \left(\frac{\Lambda_4}{m_b}\right)^{\frac{2}{23}}, \qquad \Lambda_4 = \Lambda_5 \left(\frac{m_b}{\Lambda_5}\right)^{\frac{2}{25}}, \tag{7.23}$$

$$n_f = 3 \leftrightarrow 4: \qquad \Lambda_4 = \Lambda_3 \left(\frac{\Lambda_3}{m_c}\right)^{\frac{2}{25}}, \qquad \Lambda_3 = \Lambda_4 \left(\frac{m_c}{\Lambda_4}\right)^{\frac{2}{27}}. \tag{7.24}$$

7.4 NON-PERTURBATIVE QCD

It is often mentioned that QCD only has a single free parameter, the strong coupling α_s. However, even in the perturbative region, the β-function depends explicitly on the number of quark flavours, as we have seen, and thereby also on the quark masses. Furthermore, in the non-perturbative region around or below Λ_{QCD}, the value of the perturbative coupling, as obtained, *e.g.*, from Eq. (7.20), gives little or no insight into the behaviour of the full theory. Instead, universal functions (such as parton densities, form factors, fragmentation functions, etc.), effective theories (such as the OPE, CPT or HQET [42]) or phenomenological models (such as Regge theory [47] or the string [48] and cluster [49–51] fragmentation/hadronisation models) must be used, which in turn depend on additional non-perturbative parameters whose relation to, *e.g.*, $\alpha_s(M_Z)$, is not a priori known.

For some of these questions, such as hadron masses, lattice QCD can provide important additional insight, but for multi-scale and/or time evolution contexts, the applicability of lattice methods is still severely restricted: the lattice formulation of QCD requires a Wick rotation to Euclidean space. The time coordinate can then be treated on an equal footing with the other dimensions, but intrinsically Minkowskian problems, such as the time evolution of a system, are inaccessible. The limited size of current lattices also severely constrain the scale hierarchies that it is possible to “fit” between the lattice spacing and the lattice size. We will now leave aside the non-perturbative phase of QCD and will come back to test its perturbative one in the next chapter.

CHAPTER 8

Tests of QCD Interactions

In this chapter, we will describe the experimental tests that have proven the $SU(3)_C$ structure of QCD. We will do so by initially giving an historical overview of how such a theory (as described in the previous chapter) came to be from a theoretical perspective and will eventually concentrate on its experimental verification by using data from e^+e^- experiments only. For some of content of this chapter we are indebted to Ref. [52].

8.1 TOWARDS A THEORY OF STRONG INTERACTIONS

When, in 1964, following the success of the so-called *eightfold way*, an organisational scheme for classifying hadrons proposed by Murray Gell-Mann [53] and Yuval Ne'eman [54] three years earlier, Murray Gell-Mann himself postulated *quarks* [55] and George Zweig proposed *aces* [56, 57] as the fundamental constituents of strongly interacting particles, *i.e.*, baryons and mesons, several questions immediately came to mind. What are the (very) strong forces that bind them inside the observed hadrons? What are the carriers of such forces? Are they perhaps analogous to photons? That is, just like the latter bind electrons to nuclei inside atoms, could the former bind quarks inside hadrons? But, in such a case, why do we not see individual quarks just like we see free electrons? In essence, what is the important intrinsic difference between the underlying theory of the strong interactions and QED?

DOI: 10.1201/9780429443015-8

Despite such differences, it was natural to suppose that the forces between quarks were mediated by photon-like bosons, which came to be known as gluons (as they glue quarks inside hadrons). If so, did they have spin one, like the photon? If so, did they couple to some (discrete) 'strong' charge analogous to the electric one? On the one hand, if they were Abelian vector bosons as in QED, how could they confine quarks? On the other hand, non-Abelian gauge theories were already known, yet nobody knew how to calculate reliably their dynamics. It was finally suggested that quarks carried a new quantum number called *colour* [58, 59], which could successfully explain certain puzzles such as the apparent symmetry of the wave functions of the lightest baryons and the rate for $\pi^0 \to 2\gamma$ decay. However, the postulation of colour raised additional questions. For example, did quarks of different colours all have the same EM charge and could individual coloured quarks ever be observed as free objects or would they always remain confined inside hadrons?

It was in 1968 that the results from DIS in electron-proton experiments at SLAC first became known and their importance recognised [60, 61]. Their (near) *scaling* behaviour was interpreted by James Bjorken [62,63] and Richard Feynman [64] in terms of point-like constituents within the proton, called *partons*. It was then rather natural to conclude that the partons probed by electrons though virtual photon exchange in DIS might be the quarks seen at SLAC and, indeed, this expectation was soon supported by measurements of the ratio of the longitudinal to transverse cross section [65]. However, this insight immediately raised new questions. Might the proton then contain other partons, possibly the gluons themselves, that were not probed directly by photons in DIS? And what was the origin of the bizarre dynamics that enabled quarks to resemble quasi-free point-like particles when probed at short distances in the SLAC experiments, yet confined them within hadrons, *i.e.*, when probed at large distances?

Chris Llewellyn-Smith pointed out that one could measure the total fraction of the proton momentum carried by quarks [66] and the experiments showed that this fraction was about half the total. Either the parton model was wrong or the remaining half of the proton momentum had to be carried by partons without electric charges, such as EM neutral gluons. This was the first circumstantial evidence for the existence of gluons as the carriers of the QCD interactions.

The actual existence of gluons was finally revealed in the Summer of 1978 after the PLUTO detector [67] at the e^+e^- collider DORIS (at DESY) produced the first evidence that the hadronic decays of the very narrow resonance $Y(9.46)$ could be interpreted as three-jet events generated by three gluons. Later, published analyses by the same experiment confirmed this interpretation and also hinted at the spin-1 nature of the gluon [67–69]. In the Summer of 1979, at the higher energies of the e^+e^- collider PETRA (always at DESY), again, three-jet topologies were observed and now interpreted as $q\bar{q}g$ events, eventually clearly visible also in the TASSO [70], MARK-J [71] and PLUTO experiments [72] (later, in 1980, also in JADE [73]). The spin 1 nature of the gluon was finally confirmed in 1980 by the TASSO [74] and PLUTO [75] experiments. (See Refs. [52, 76] for reviews on the history of the gluon.)

8.2 MAKING OF QCD

The measurement of α_s was made possible by a variety of experiments through a collection of observables. Amongst the latter though, those which have led to a substantial progress in the understanding of the perturbative phase of QCD are connected to the study of the cross section of the process $e^+e^- \to$ hadrons. The attentive reader would however say that this is somewhat of a throw-back, as we have just discussed that the hadronic phase of QCD is non-perturbative. Indeed, QCD Feynman rules tell us only about partons living at small distances while hadron formation is a long distance process. So the question then arises: how do we calculate $e^+e^- \to$ hadrons? How can we possibly ignore the vast amount of physics which develops between partons and hadrons, *i.e.*, parton shower (a leading logarithmic description of the QCD evolution following the hard scattering) and fragmentation/hadronisation? A schematic view of this is given in Fig. 8.1.

Besides, the final state 'hadrons' is not really well defined. In fact, each event has a different hadronic final state. However, none of all this is actually a concern, as we really aim at integrating out all of the unknowns, which indeed includes summing over all possible hadronic final states. We are really looking at the fully inclusive cross section to produce any hadronic final state in e^+e^- annihilations.

In order to establish a link between partons and hadrons, we will exploit symmetries of the aforementioned cross section. Let us start by writing the

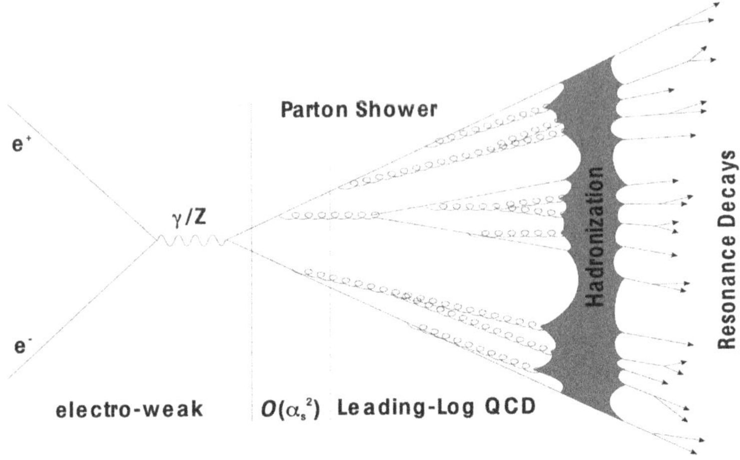

Figure 8.1 The structure of the process $e^+e^- \to$ hadrons from the perturbative (partons) to non-perturbative (hadrons) phase.

ME to produce n hadrons $h_1, ..., h_n$ as follows:

$$\mathcal{M} \sim \{\bar{v}(p_{e^+})e\gamma_\mu u(p_{e^-})\} \ \frac{-g^{\mu\nu}}{q^2} \ T_\nu(n, q, \{p_{h_1} \ldots, p_{h_n}\}), \tag{8.1}$$

where $p_{e^\pm}$ are the electron/positron momenta, $q^2 = q_\mu \cdot q^\mu$ (the square of the collision energy) and $p_{h_1,\ldots,h_n}$ are the hadron momenta, with T_ν being a parameterisation of the unknown part arising in the final state of any event. In a sense, this tensor parameterises our *ignorance* of the dynamics taking us from the partonic to the hadronic stage.

If we now compute the total cross section, using well-known properties of the Dirac matrices, we obtain

$$\begin{aligned}\sigma = &\frac{1}{2s}\,\frac{1}{4}\,\frac{e^2}{s^2}\mathrm{Tr}(\not{p}_{e^+}\gamma^\mu \ \not{p}_{e^-}\gamma^\nu) \\ &\times \sum_n \int d\mathrm{PS}_n \ T_\mu(n, q, \{p_{h_1}, \ldots, p_{h_n}\}) \ T^*_\nu(n, q, \{p_{h_1}, \ldots, p_{h_n}\}).\end{aligned} \tag{8.2}$$

Now, let us define

$$H_{\mu\nu}(q) \equiv \sum_n \int dPS_n \ T_\mu \ T^*_\nu \tag{8.3}$$

and impose Lorentz covariance, *i.e.*,

$$H_{\mu\nu} = Ag_{\mu\nu} + Bq_\mu q_\nu, \tag{8.4}$$

wherein it is clear that A and B are functions only of q^2. Let us further impose gauge invariance, *i.e.*:

$$q^\mu H_{\mu\nu} = q^\nu H_{\mu\nu} = 0, \tag{8.5}$$

which in turn implies

$$A = -q^2 B. \tag{8.6}$$

Hence, it follows that

$$\sigma = \frac{e^2}{2s} B(s) \tag{8.7}$$

with $B(s)$ dimensionless. This finally gives the fundamental prediction:

$$R \equiv R(e^+e^-) = \frac{\sigma(e^+e^- \to \text{hadrons})}{\sigma(e^+e^- \to \mu^+\mu^-)} = \text{constant}, \tag{8.8}$$

without actually knowing anything about parton-to-hadron dynamics.

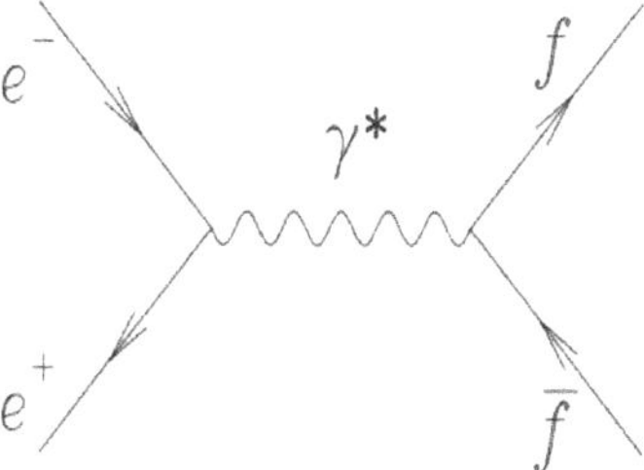

Figure 8.2 The LO diagram contributing at parton level to the $e^+e^- \to$ hadrons cross section (when $f = q$).

Let us now try to understand what this constant is. Let us picture our $e^+e^- \to$ hadrons cross section at LO and limit ourselves to the case of photon exchange in Fig. 8.1 (which is in fact an accurate description for $s \ll M_Z^2$). By using the QED Feynman rules, it is straightforward to obtain from Fig. 8.2 that (using the 0 subscript to signify the LO result)

$$\sigma_0 = \frac{4\pi\alpha^2}{3s} N_C \sum_q e_q^2 \ (\text{if} f = q), \tag{8.9}$$

i.e., if the fermion f is a quark q (as necessary to initiate an hadronic final state) one also has that

$$R_0 \equiv \frac{\sigma_0(e^+e^- \to \text{hadrons})}{\sigma_0(e^+e^- \to \mu^+\mu^-)} = N_C \sum_q e_q^2, \tag{8.10}$$

where e_Q is the electric charge of q and N_C the number of colours. Here, the summation over the number of quark flavours is extended to q for which $\sqrt{s} > 2m_q$, so that one would expect R_0 to be a sequence of step functions raising at $\sqrt{s} = 2m_q$ for each quark flavour q. This is well exemplified in Fig. 8.3, which collects a variety of R measurements at various e^+e^- experiments running a different energies $E_{\text{cm}} \equiv \sqrt{s}$. Herein, the two steps at the J/ψ (*i.e.*, $E_{\text{cm}} \approx 2m_c$) [19, 20] and Y/Υ (*i.e.*, $E_{\text{cm}} \approx 2m_b$) [22] hadronic masses are clearly visible, enabling one to extract the universal (*i.e.*, flavour independent) value of N_C: *e.g.*, at the PETRA collider ($\sqrt{s} = 34$), just above the $2m_b$ threshold, $R = 3.88 \pm 0.03$ was measured, well consistent with the theoretical prediction of $R = \frac{11}{3}$, obtained for $N_C = 3$. In short, the number of colours defining the $SU(3)_C$ structure of QCD was measured experimentally and indeed proved to be the same for every quark flavour, thereby also confirming the flavour independence of QCD interactions (as well as enabling one to measure the EM charge e_q of any newly found quark).

This beautiful result could have been repeated for every new quarks eventually discovered. It turned out, though, that the sixth quark of the SM, the top one, is rather heavy, some 173 GeV, so that the $t\bar{t}$ threshold has never been in reach of e^+e^- machines that we have constructed so far, as the highest energy reached to date is about 209 GeV, as attained by the LEP collider towards the end of the nineties. Another complication is that, as soon as $\sqrt{s}$ becomes comparable to M_Z, the contribution due to Z mediation becomes comparable to the one due to the γ^* channel (a hint of the Z contribution can already be seen in Fig. 8.3 for $E_{\text{cm}} \approx 40$ GeV) hence, the simple expression in Eq. (8.9) is no longer applicable. However, the corresponding LO theoretical prediction at $\sqrt{s} = M_Z$ is well known,

$$R = N_C \frac{\sum_q \mathcal{A}_q}{\mathcal{A}_\mu} = 20.095, \qquad \mathcal{A}_f = v_f^2 + a_f^2, \tag{8.11}$$

and, in fact, account for a dependence upon the vector and axial Z couplings, v_q and a_q, respectively, which can be used to verify the EW quantum

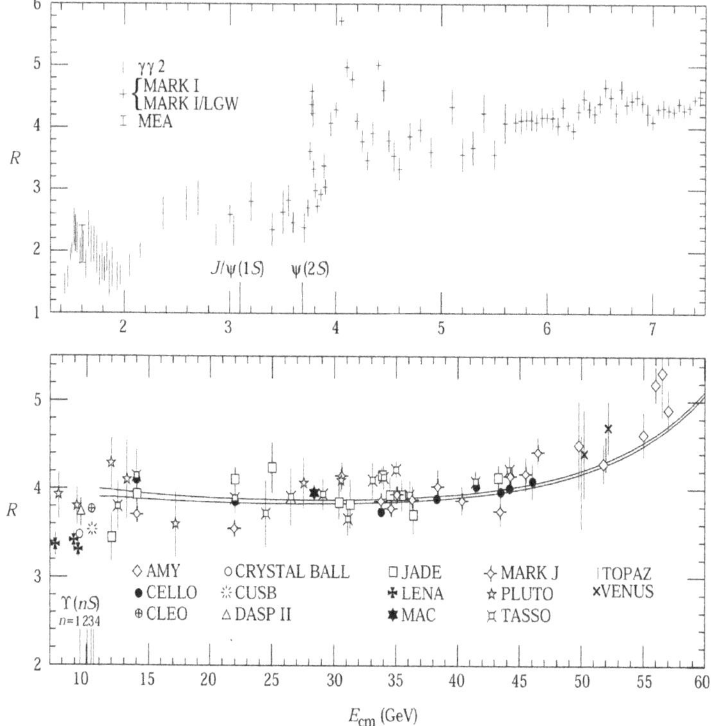

Figure 8.3 The function R given in the text as measured by several electron-positron experiments at different energies (adapted from [77]).

numbers of any q state found. (We will come back to this in a later chapter.) Remarkably, this is very consistent with the LEP average measurement of $R = 20.775 \pm 0.027$. However, as soon as we depart from the peak position $\sqrt{s} = M_Z$, in either direction, interference effects between the photon and Z propagators become important (and the expression for R far more complicated). The characteristic shape of R appearing around M_Z in the presence of all such effects can be seen in Fig. 8.4. Herein, it can finally be seen that, asymptotically, *i.e.*, for $\sqrt{s} \approx 2m_t$, one should also be able to see the (approximate) step function induced in the R dependence by the top quark, which is indeed within the scope of future e^+e^- colliders.

So far so good. However, our attentive reader may now well say: hang on, but all this only tells me that there can me many quarks, each coming with N_C colours (and the EW quantum numbers predicted by the SM), but nothing

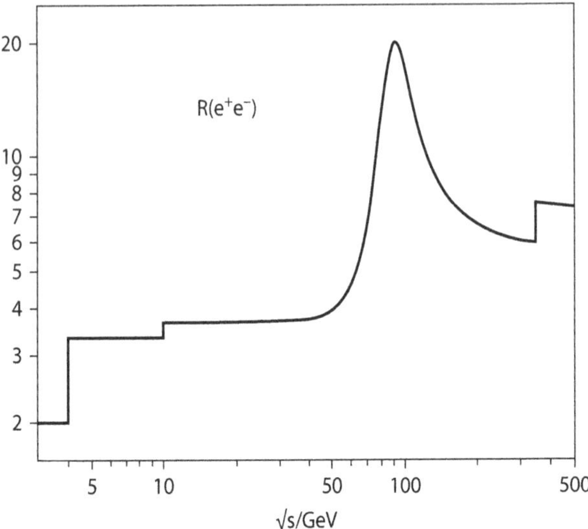

Figure 8.4 The function R given in the text as predicted by the SM at LO as a function of the collider energy.

really about the dynamics of QCD, *i.e.*, about the presence of gluons and their interaction, in turn leading to the characteristic behaviour of α_s that we have encountered in the previous chapter. Quite rightly, in order to learn about the latter, we need to move on and study $e^+e^- \to$ hadrons at NLO.

Now, α_s is the largest coupling in the SM (at least at the energy scale we are presently able to probe), so we should indeed expect QCD corrections being largest. Through NLO in α_s, the diagrams intervening in the computation of R are given in Fig. 8.5.

They enter the calculation as follows. There are virtual (gluon) corrections which are obtained by interfering the tree-level diagrams with one-loop ones in Fig. 8.5(a) as well as real (gluon) corrections which are obtained by squaring the tree-level diagrams in Fig. 8.5(b). The former, being an interference, can be negative. The latter, being a square, are definite positive. Let us look at them in turn, starting with the real QCD corrections, *i.e.*, those involving the emission of a real gluon.

In order to do so, we need to introduce conveniently a 3-body phase space, which we do as follows:

$$d\Phi_3 \propto d\alpha\, d\beta\, d\gamma\, dx_1\, dx_2, \tag{8.12}$$

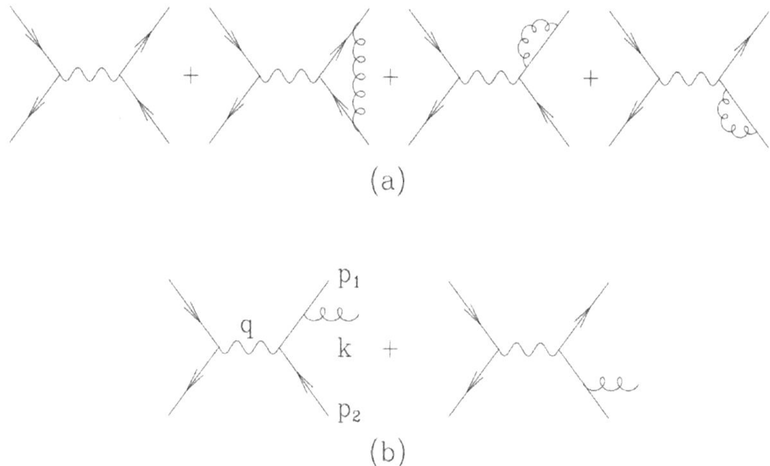

Figure 8.5 The NLO diagrams contributing to the $e^+e^- \to$ hadrons cross section.

where α, β and γ are the Euler angles while $x_1 = 2E_q/\sqrt{s}$ and $x_2 = 2E_{\bar{q}}/\sqrt{s}$ are energy fractions of final-state quark and antiquark, respectively. Upon applying Feynman rules and integrating over the Euler angles, one obtains

$$\sigma^{q\bar{q}g} = 3\sigma_0 C_F \frac{\alpha_s}{2\pi} \int dx_1 dx_2 \frac{x_1^2 + x_2^2}{(1-x_1)(1-x_2)}, \tag{8.13}$$

with integration region $0 \leq x_1, x_2 \leq 1$. Remarkably, such an integral is divergent at $x_1, x_2 = 1$, as we can rewrite the two terms entering in the denominator above as follows

$$1 - x_1 = x_2 x_3 (1 - \cos\theta_{qg})/2, \tag{8.14}$$

$$1 - x_2 = x_1 x_3 (1 - \cos\theta_{\bar{q}g})/2, \tag{8.15}$$

where $x_3 = 2E_g/\sqrt{s}$ (with E_g the gluon energy) and $\theta_{ig}(i = 1, 2)$ are the angles between gluon and (anti)quarks.

There are two types of divergences:

- collinear divergence, when $\theta_{qg} \to 0$ or $\theta_{\bar{q}g} \to 0$;
- soft divergence, when $E_g \to 0$.

We cumulatively call these Infra-Red (IR) divergences and we note that the two types can occur separately or together. Such singularities simply indicate the breakdown of perturbation theory when energy scales approach $\Lambda_{\rm QCD}$, *i.e.*, the Landau pole. As troubling as they may appear, fortunately, such collinear and soft regions do not make important contributions to the total cross section, simply because the ensuing divergences cancel against similar ones emerging from virtual gluon emissions.

Let us see how in some detail. First, let us perform the above (finite) integral over x_1 and x_2 by using, *e.g.*, what is called *dimensional regularisation*[1]. That is, rather than integrating in the customary $D = 4$ dimensions of space-time, we integrate over $D = 4 - 2\epsilon$ dimensions, so that we obtain

$$\sigma^{q\bar{q}g} = 3\sigma_0 C_F \frac{\alpha_s}{2\pi} H(\epsilon) \int dx_1 dx_2 \frac{(1-\epsilon)(x_1^2 + x_2^2) + 2\epsilon(1-x_3)}{(1-x_3)^\epsilon[(1-x_1)(1-x_2)]^{1+\epsilon}}, \quad (8.16)$$

where $H(\epsilon) = \frac{3(1-\epsilon)(4\pi)^{2\epsilon}}{(3-2\epsilon)\Gamma(2-2\epsilon)} = 1 + \mathcal{O}(\epsilon)$. Hence,

$$\sigma^{q\bar{q}g} = 3\sigma_0 C_F \frac{\alpha_s}{2\pi} H(\epsilon) \left[\frac{2}{\epsilon^2} + \frac{3}{\epsilon} + \frac{19}{2} - \pi^2 + \mathcal{O}(\epsilon)\right]. \quad (8.17)$$

That is, soft and collinear divergences are *regulated*, appearing as *poles* at $D = 4$ ($\epsilon = 0$). Second, it can be shown that the diagrams embedding virtual gluon exchange lead to a contribution, integrated over a 2-body phase space of $D = 4 - 2\epsilon$ dimensions, of the following form:

$$\sigma^{q\bar{q}} = 3\sigma_0 \left\{1 + C_F \frac{\alpha_s}{2\pi} H(\epsilon) \left[-\frac{2}{\epsilon^2} - \frac{3}{\epsilon} - 8 + \pi^2 + \mathcal{O}(\epsilon)\right]\right\}. \quad (8.18)$$

Therefore, by adding real and virtual corrections, the IR poles cancel and the result is finite as $\epsilon \to 0$:

$$R = R_0 \left\{1 + 1.045 \left(\frac{\alpha_s}{\pi}\right) + \mathcal{O}(\alpha_s^2)\right\}, \quad (8.19)$$

where R_0 is the LO term previously computed and the coefficient in front of α_s is computed at $q^2 = M_Z^2$. We thus conclude that R is an *IR safe* quantity, being *finite* and *regularisation scheme-independent*, as the aforementioned cancellations occur to any order in α_s. Nowadays, R is known to NNNLO in QCD,

[1]Other *regularisation schemes* are available, *e.g.*, using a finite g mass $m_g \equiv \epsilon s$, which would lead to the same final result, yet the intermediate steps would be non-gauge invariant.

through the formula

$$R = R_0 \left\{ 1 + 1.045 \left(\frac{\alpha_s}{\pi} \right) + 0.94 \left(\frac{\alpha_s}{\pi} \right)^2 - 15 \left(\frac{\alpha_s}{\pi} \right)^3 + \mathcal{O}(\alpha_s^4) \right\}, \quad (8.20)$$

where the constant coefficients are computed at M_Z. It is precisely this formula that has enabled the extraction of α_s from the measurements at e^+e^- colliders listed in the previous chapter.

8.3 CONFIRMATION OF QCD

As mentioned, the assumption that $SU(3)_C$ is behind QCD implies the existence of 8 mediators for the strong interactions, called gluons. These are massless (as QCD is unbroken), coloured (hence, self-interacting as QCD is non-Abelian) and carry spin 1 (just like the mediators of the EW force). Soon after the postulation of quarks, it was suggested that they interact via such gluons, but direct experimental evidence was lacking for over a decade. In 1976, Mary Gaillard, Graham Ross and John Ellis suggested searching for gluons via 3-jet events induced by gluon bremsstrahlung in e^+e^- collisions (see Fig. 8.5b). Thanks to the robustness of the QCD predictions, as detailed above, which was put on firm grounds by a 1977 paper from Sterman and Weinberg [78], following such a suggestion, the gluon was discovered at DESY in 1979 by TASSO [79] and the other experiments at the PETRA [71] collider by direct confrontation of QCD predictions with data. Indeed, further investigations of data, specifically of the angular distribution of the gluon jets with respect to the quark ones, confirmed its spin-1 nature.

However, it was not till the advent of the LEP collider that $SU(3)_C$ of QCD was fully proven by experiment. The postulated gauge structure of the QCD interactions can only be unmistakenly confirmed once the Casimirs of the fundamental (C_F) and adjoint representation (C_A) are measured. As shown in Tab. 1.1, they can be expressed in terms of the Gell-Mann matrices t^a and structure constants captured by the (squared) diagram corresponding to C_A, as it signals the self-interacting properties of gluons.

In e^+e^- annihilations, such interactions emerge in 4-parton final states, eventually generating 4-jet events, via the Feynman diagrams depicted in Fig. 8.6(c). However, the latter cannot really be separated from those in Fig. 8.6(a)–(b), owing to the gauge invariance of QCD, which only carry gluon

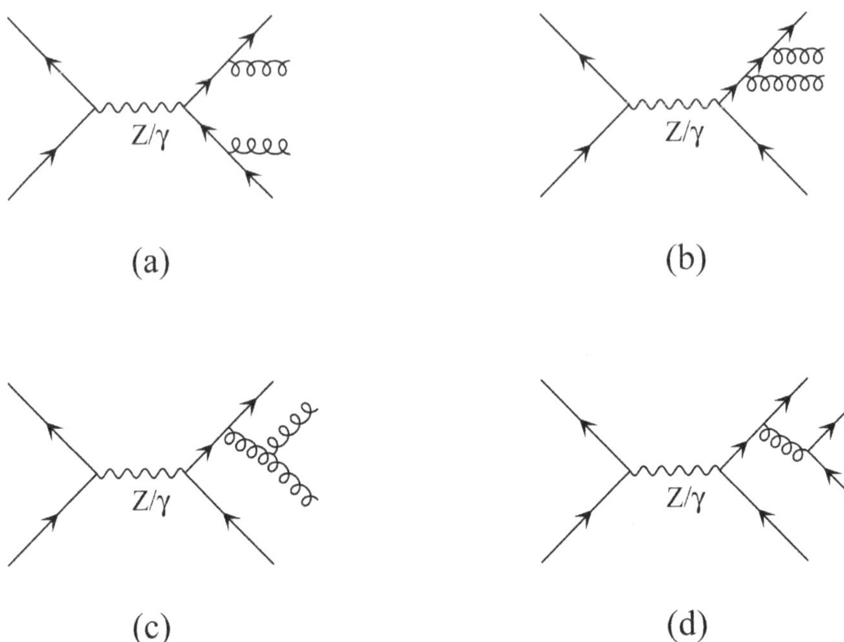

Figure 8.6 Feynman diagrams leading to 4-jet events in e^+e^- annihilations to $q\bar{q}gg$ and $q\bar{q}Q\bar{Q}$: (a) double gluon bremsstrahlung from (different) (anti)quarks; (b) double gluon bremsstrahlung from the (same) (anti)quark; (c) triple-gluon vertex diagram; (d) four-quark production.

interactions which are Abelian in nature, like in QED (*i.e.*, double photon bremstrahlung). In fact, recalling that (anti)quarks and gluons do not appear as free particles, rather they produce jets which cannot unambiguously be identified with the original parton, one also has to account for the diagrams in Fig. 8.6(d), which also have an Abelian structure. It is however possible to fit experimental data made up by 4-jet events to theoretical predictions expressed in terms of C_A and C_F (see Ref. [80] for the actual expressions) and produce results such as those obtained by the four LEP collaborations in Fig. 8.7 (also including other experimental observables, such as event shapes [81]), highlighting the success of $SU(3)_C$ in explaining QCD interactions against competing theories.

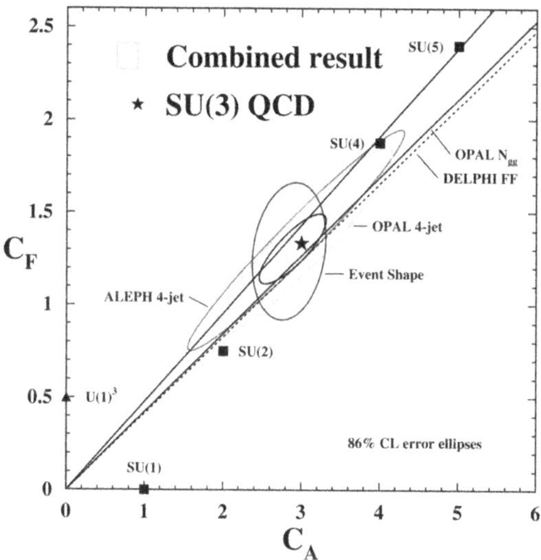

Figure 8.7 LEP measurements of C_A and C_F. (Figure from Ref. [82].)

CHAPTER 9

Higgs Phenomenology

In this chapter, we set the stage for the discovery of the SM Higgs boson, by recapping what was the theoretical situation ahead of the LHC starting operations, which led to the discovery of a SM-like Higgs boson in July 2012. We will review the SM Higgs boson properties (mass and couplings) as well as its production and decay rates as expected at the time. Specifically, at this point of our narration, we will dismiss the present knowledge that its mass is now measured at 125 GeV, rather we will keep it as a free parameter (which it is, in fact, in the SM, as we have previously seen). This is particularly important as it will make us realise that, once again, as Albert Einstein famously put it[1]: 'Subtle is the Lord, but malicious He is not'. In fact, had the mass of the Higgs boson eventually discovered at CERN been different, it may have been easier to discover it or else absolutely impossible, as we shall remark upon below. As the quest for the elusive Higgs boson lasted for nearly 40 years, a significant amount of technical development went into predicting its possible signals, so we will spend some time on this too. (This chapter is largely based on Ref. [83], from which some of the plots are taken.)

9.1 HIGGS MASS AND COUPLINGS

As previously explained, in the SM, the mass (and couplings) of the Higgs boson are arbitrary parameters. However, there are ways to constrain them, by requiring self-consistency of the theory. Specifically, unitarity of EW interactions requires the existence of a scalar Higgs field which couples to other

[1]Remark made during Einstein's first visit to Princeton University (April 1921) as quoted in *Einstein* (1973) by R. W. Clark, Chapter 14.

DOI: 10.1201/9780429443015-9

$$V \quad \Longrightarrow \quad + \quad + \quad + \cdots$$

$$\frac{1}{q^2} \quad \Rightarrow \quad \frac{1}{q^2} + \sum_j \frac{1}{q^2}\left[\left(\frac{g_W v}{\sqrt{2}}\right)^2 \frac{1}{q^2}\right]^j = \frac{1}{q^2 - M_V^2} \; : \; M_V^2 = g_W^2 \frac{v^2}{4}$$

Figure 9.1 Schematic representation of boson mass generation.

particles proportional to their mass. A convenient way of seeing how this dynamics emerges in the SM is by referring to Figs. 9.1 and 9.2, wherein it is made intuitively evident how the masses of (initially massless) gauge bosons V and fermions f build up by (infinitely) repeated interactions with the background Higgs field. The mass of the Higgs boson itself is extracted from the curvature of the scalar potential V, as $M_H^2 = \lambda v^2$. This makes it clear that the SM Higgs boson mass cannot be predicted since the quartic coupling λ is unknown. Nevertheless, restrictive bounds on M_H follow from the hypothetical assumptions on the energy scale Λ up to which SM is valid before some NP emerges.

It all starts from quantum fluctuations introducing an energy dependence, so that $\lambda \equiv \lambda(\mu^2)$. This happens through the following diagrams (at one-loop level):

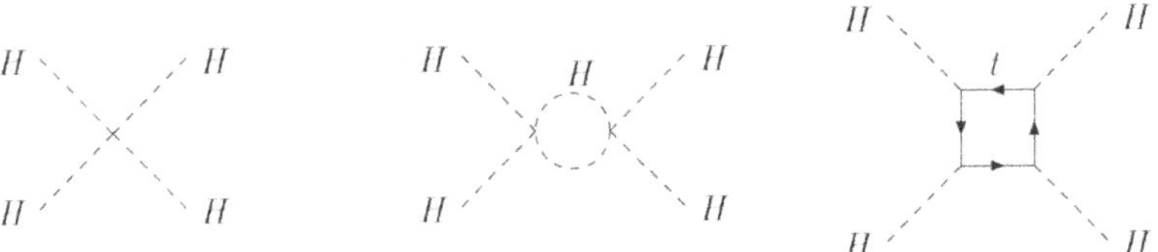

Such a $\lambda(\mu)$ running can conveniently be described through its RGE

$$\frac{d\lambda}{d\ln \mu^2} = \frac{3}{8\pi^2}[\lambda^2 + \lambda g_t^2 - g_t^4], \tag{9.1}$$

with $\lambda(v^2) = M_H^2/v^2$ and $g_t(v^2) = \sqrt{2}m_t/v$, wherein each of the terms is in a one-to-one correspondence with the diagrams above. For moderate m_t and large M_H, one obtains

$$d\lambda/d\ln \mu^2 \sim \lambda^2, \tag{9.2}$$

$$f \longrightarrow \quad \Longrightarrow \quad \longrightarrow \quad + \quad \times H \quad + \quad \times \times \quad + \cdots$$

$$\frac{1}{\not{q}} \quad \Rightarrow \quad \frac{1}{\not{q}} + \sum_j \frac{1}{\not{q}} \left[\frac{g_f v}{\sqrt{2}} \frac{1}{\not{q}} \right]^j = \frac{1}{\not{q} - m_f} \quad : m_f = g_f \frac{v}{\sqrt{2}}$$

Figure 9.2 Schematic representation of fermion mass generation.

so that an approximate solution can be found, as follows:

$$\lambda(\mu^2) = \frac{\lambda(v^2)}{1 - \frac{3\lambda(v^2)}{8\pi^2} \ln \frac{\mu^2}{v^2}}, \tag{9.3}$$

thus revealing a singularity at the Landau pole (this is not dissimilar from the QCD one), *i.e.*, when $\ln \frac{\mu^2}{v^2} = \frac{8\pi^2}{3\lambda(v^2)}$. The requirement of the SM remaining perturbative up to a scale Λ, such that $\lambda(\Lambda) < \infty$, can be translated into a rough upper bound on M_H (so-called triviality bound), as follows

$$M_H^2 \lesssim \frac{8\pi^2 v^2}{3 \ln (\Lambda^2/v^2)}. \tag{9.4}$$

Notice that such an upper bound on M_H depends on the cut-off value Λ.

Furthermore, upon noting that top-loop corrections reduce λ for increasing m_t, so that λ becomes negative if m_t is too large, a lower bound on M_H can be derived. In other words, the self-energy potential would become deeply negative and the ground state would no longer be stable. Thus, to avoid instability, M_H must exceed a minimal value for a given m_t to balance the negative top-quark contribution. In practice, for small λ (*i.e.*, small M_H), the last term in Eq. (9.1) dominates and one obtains

$$\lambda(\Lambda^2) = \lambda(v^2) - \frac{3}{8\pi^2} g_t^2 \ln \frac{\Lambda^2}{v^2}, \tag{9.5}$$

from where, upon requiring $\lambda(\Lambda^2) > 0$, a rough lower bound on the Higgs mass is obtained (so-called vacuum stability bound):

$$M_H^2 \gtrsim \frac{3v^2}{8\pi^2} g_t^2 \frac{\Lambda^2}{v^2}. \tag{9.6}$$

Hence, also the lower bound on M_H depends on the cut-off value Λ.

The fact that both lower and upper bounds on the Higgs boson mass depend on Λ, which is essentially the cut-off beyond which the SM is no longer valid; hence the introduction of some NP is required, allows one to draw a plot like in Fig. 9.3. Herein, the lower and upper bands, which thickness incorporates the systematic error stemming from the theory, which is due to the fact that one can compute the RGE for $\lambda(\mu^2)$ only to a finite order in the perturbative expansion, enable to extract one key information relating to the scale of NP, Λ, where the SM will cease to be valid, as a function of the Higgs mass, M_H. Or conversely, to read the value of M_H for which the SM can be a self-consistent framework up to energies where a GUT ($\Lambda = 10^{19}$ GeV) is required, also incorporating gravity (which will become as strong as the other forces at the Planck mass, $\Lambda = M_{\text{Planck}} = 10^{16}$ GeV).

Taking $m_t = 173$ GeV and by using the mean value of the two predictions in the figure, one can tabulate the values for the Higgs mass and the scale of NP as given in Tab. 9.1.

Λ	M_H
1 TeV	$M_H \approx 55$ or 700 GeV
10^{19} GeV	130 GeV $\lesssim M_H \lesssim$ 190 GeV

Table 9.1 Scale of NP versus the M_H range requiring it.

This makes the intriguing point that, *e.g.*, if the SM were valid up to the GUT scale, then the value of the Higgs mass should fall in the range 130 GeV $\lesssim M_H \lesssim$ 190 GeV. Conversely, an observation of M_H outside this range would demand NP below the GUT scale. As an extreme example, a Higgs mass around 55 or 700 GeV would have required some NP already at the TeV scale, which is the energy realm of the LHC. We leave it to the attentive reader to ponder about Einstein's remark reported at the beginning of this chapter when recalling that the CERN machine has finally discovered a Higgs boson, with properties similar to the SM one, with a mass of approximately 125 GeV, *i.e.*, right at the border of the mass region which might make the SM a self-consistent framework without the need of NP, ever. Or might it not? Before addressing this question, though, let us review how we prepared for such a discovery.

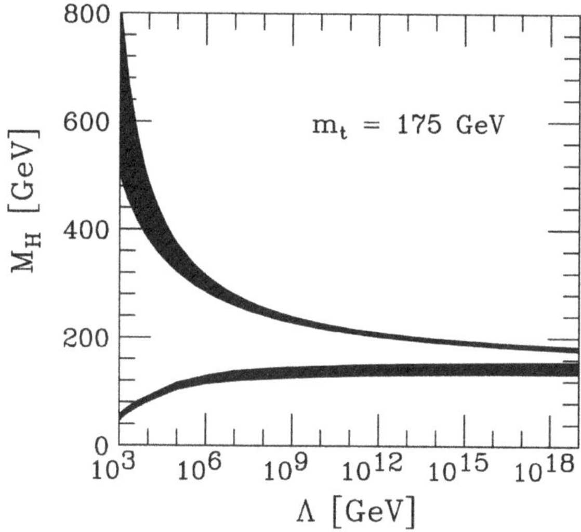

Figure 9.3 The upper and lower limits on M_H (the Higgs mass) as a function of Λ (the scale of NP).

9.2 HIGGS DECAY BRANCHING RATIOS

The BRs of the SM Higgs boson have been studied in many papers over the years. A useful compilation of the early works on this subject can be found in Refs. [84, 85], where all the most relevant formulae for on-shell decays were summarised. Higher-order corrections to most of the decay processes have also been computed (for some reviews, see [86–88] and references therein) as well as the rates for the off-shell decays $H \to W^*W^*, Z^*Z^*$ [89–91], $H \to Z^*\gamma$ [92] and $H \to t^*\bar{t}^*$ [93, 94].

Large QCD corrections to the SM Higgs partial widths enter into the following channels: heavy quark pairs [95,96], $Z\gamma$, $\gamma\gamma$ [97,98] and gg [99,100]. In fact, the QCD corrections to the top loops in $Z\gamma$, $\gamma\gamma$ and gg decays computed for $M_H \ll 2m_t$ are actually accurate enough to be used for any m_t value[2].

The bulk of the QCD corrections to $H \to q\bar{q}$ can be absorbed into a 'running' quark mass $m_q(\mu)$, evaluated at the energy scale $\mu = M_H$. The importance of this effect for the case $q = b$ was discussed in Ref. [101]. There is, however, a slight subtlety concerning $t\bar{t}$ decays [94]. For $H \to q\bar{q}$ decays

[2]Numerical results valid for any value of the ratio $\tau = M_H^2/4m_t^2$ can be found in Ref. [100].

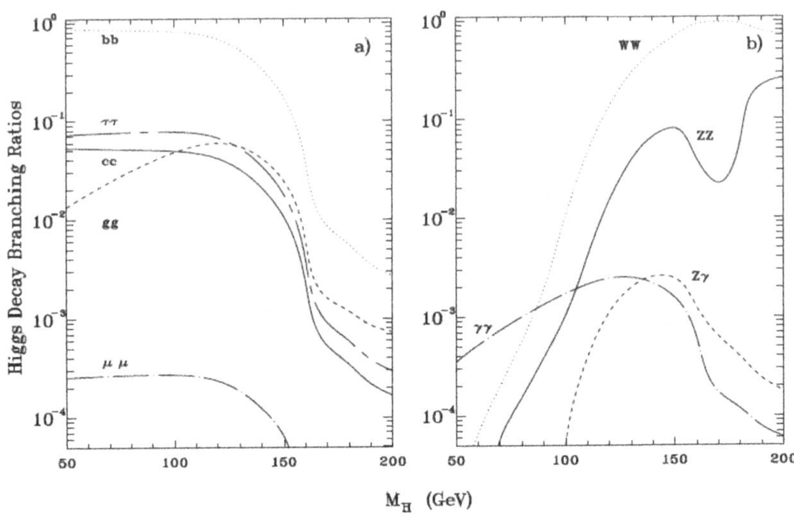

Figure 9.4 BRs of the SM Higgs boson in the mass range 50 GeV $< M_H <$ 200 GeV, for the decay modes: (left) $b\bar{b}$, $c\bar{c}$, $\tau^+\tau^-$, $\mu^+\mu^-$ and gg; (right) W^+W^-, ZZ, $\gamma\gamma$ and $Z\gamma$.

involving light quarks ($q = s, c, b$), the use of the running quark mass $m_q(\mu = M_H)$ takes into account large logarithmic corrections at higher orders in QCD perturbation theory and so in principle one could imagine using the same procedure for $H \to t\bar{t}$, at least in the limit $M_H \gg m_t$. In practice, however, one is interested only in the region $M_H/m_t \sim O(1)$. In the case of the top quark loop mediated decay $H \to gg$, it is well known that the higher-order QCD corrections are minimised if the quark mass is defined at the pole of the propagator, *i.e.*, $m_t(\mu = m_t)$. To be consistent, therefore, one should use the *same* top mass $m_t(\mu = m_t)$ in the decay width for $H \to t\bar{t}$ (and $H \to Z\gamma, \gamma\gamma$ as well). For the light quark loop contributions to the $H \to gg(Z\gamma, \gamma\gamma)$ decay widths we use pole(running) masses (defined at the scale $\mu = M_H$) [102]. The results on the Higgs decay rates are summarised in Figs. 9.4 and 9.5.

Fig. 9.4 shows the BRs, for $M_H \leq 200$ GeV, the so-called intermediate mass range, for the channels: (left) $b\bar{b}$, $c\bar{c}$, $\tau^+\tau^-$, $\mu^+\mu^-$ and gg as well as (right) W^+W^-, ZZ, $\gamma\gamma$ and $Z\gamma$. Fig. 9.5 (left) shows the Higgs BRs for 200 GeV $\leq M_H \leq$ 1000 GeV, the so-called heavy mass range. The inclusion here of below-threshold $H \to t^*\bar{t}^*$ decays does not give any observable effect, since this channel is heavily suppressed by the W^+W^- and ZZ decays.

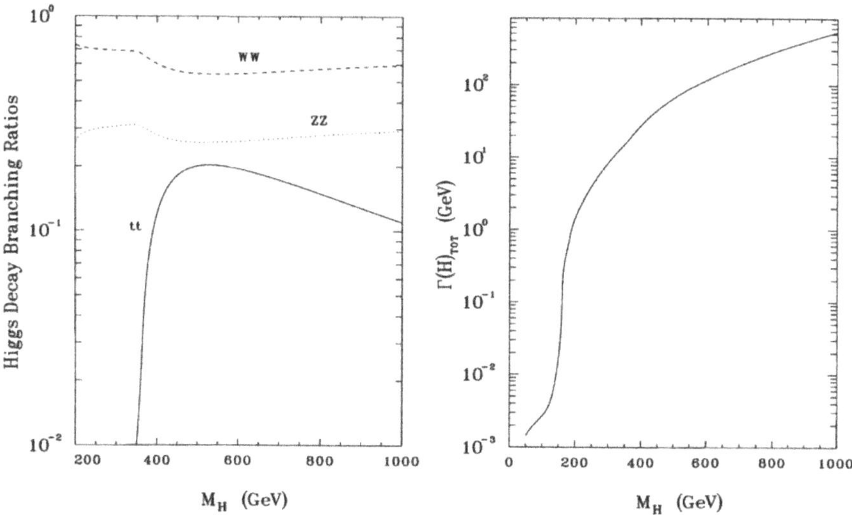

Figure 9.5 (Left) BRs of the $\mathcal{SM}$ Higgs boson in the mass range 200 GeV $< M_H <$ 1 TeV, for the decay modes: W^+W^-, ZZ and $t\bar{t}$. (Right) Total width of the $\mathcal{SM}$ Higgs as a function of the mass in the range 50 GeV $< M_H <$ 1 TeV.

For completeness, we show in Fig. 9.5 (right) the total SM Higgs decay width over the range 50 GeV $\leq M_H \leq$ 1000 GeV. A remarkable aspect of this plot is that, before the W^+W^- and ZZ channels open up, the Higgs boson is very narrow, with Γ_H being at the MeV level, which makes it clear that, whichever channels are used for its detection, the detector resolution will never be able to resolve it as a Breit-Wigner resonance at the LHC (as the best value for it is never smaller than a GeV or so).

9.3 HIGGS PRODUCTION CROSS SECTIONS

There are only a few Higgs production mechanisms which lead to detectable cross sections at the LHC. Each of them makes use of the preference of the SM Higgs to couple to heavy particles: either massive vector bosons ($W^\pm$ and Z) or massive quarks (especially t-quarks). They are (see Fig. 9.6):

(a) gluon-gluon fusion [103],

(b) W^+W^- and ZZ fusion [104],

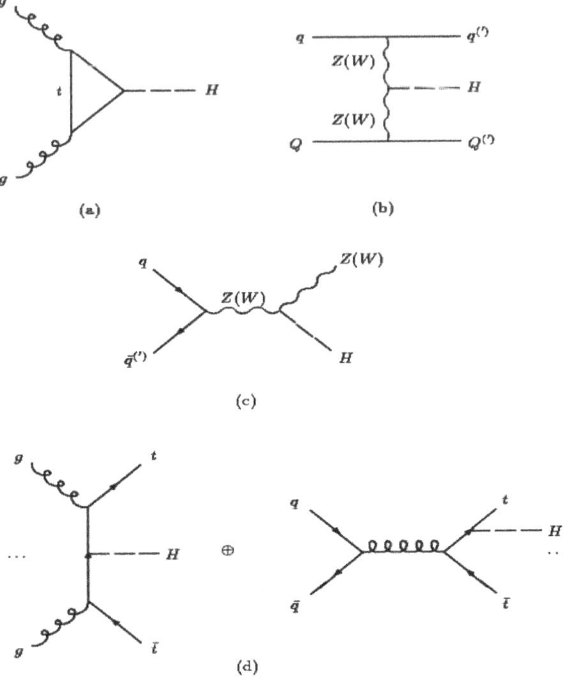

Figure 9.6 Representative Feynman diagrams describing the main mechanisms of Higgs production at the LHC: (a) gluon-gluon fusion; (b) W^+W^- or ZZ fusion (also called VBF); (c) associated production with $W^\pm$ or Z (also called HS); (d) associated production with top quark pairs.

(c) associated production with $W^\pm$ and Z bosons [105, 106],

(d) associated production with $t\bar{t}$ pairs [107, 108].

A complete review on the early literature on pp collider SM Higgs boson phenomenology, based on these production mechanisms, can be found in Refs. [84, 85, 94, 101].

The NLO QCD corrections are known for all processes (a)–(d) and included in our forthcoming plots. By far the most important of these are the corrections to the gluon fusion process (a) which have been calculated first in Ref. [109]. In the limit where the Higgs mass is far below the $2m_t$ threshold, these corrections are calculable analytically [110–112]. In fact, it turns out that the analytic

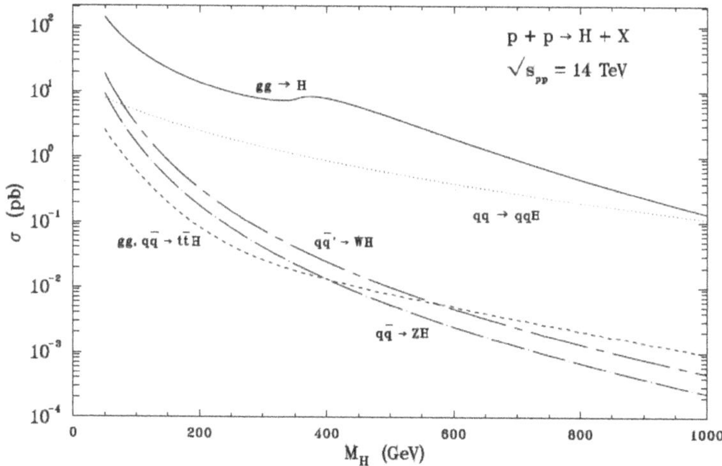

Figure 9.7 Total cross sections for H production at the LHC as a function of the Higgs mass M_H, as given by the four production mechanisms illustrated in the text at $\sqrt{s_{pp}} = 14$ TeV.

result, as mentioned, is a good approximation over the complete M_H range and so we have used it in our results [102, 113].

Overall, the NLO correction increases the leading-order result by a factor of about 2, when the normalisation and factorisation scales are set equal to $\mu = M_H$. This 'k-factor' can be traced to a large constant piece in the NLO correction [109],

$$k \approx 1 + \frac{\alpha_s(\mu = M_H)}{\pi} \left[\pi^2 + \frac{11}{2}\right]. \tag{9.7}$$

Such a large k-factor usually implies a non-negligible scale dependence of the theoretical cross section. Hence, NNLO results were eventually required to obtain a well behaved QCD expansion for the cross section of the gluon fusion process (see Ref. [114] for the latest developments in this respect).

The NLO corrections to the VV fusion [115] and VH [115] production cross sections ($V = W^\pm, Z$) are quite small, increasing the total cross sections by no more than $\approx 10\%$ and $\approx 20\%$, respectively. Note that for the former, according to Ref. [115], a factorisation scale set as $\mu^2 = -q_V^2$ is appropriate, where q_V^μ is the four-momentum of the virtual $V = W, Z$ boson. For the latter it is probably best to use $\mu^2 = \hat{s}_{pp} \approx (p_V + p_H)^2$, though the scale dependence here is very mild. The corresponding QCD corrections for the $t\bar{t}H$ mechanism were computed in Refs. [116, 117] (we use here $\mu^2 = \hat{s}_{pp}$). A detailed review of NLO QCD corrections to all Higgs cross sections can be found in Ref. [88] while, again, we refer the reader to [114] for the latest developments.

Results for the production cross sections are given in Fig. 9.7, for the representative LHC energy of 14 TeV. The pattern of the various curves is very clear: the gluon fusion mechanism is dominant over all the Higgs mass range, followed by W^+W^-/ZZ fusion which becomes comparable in magnitude to gluon-gluon fusion for very large Higgs masses. The cross sections of the other production mechanisms (WH, ZH and $t\bar{t}H$) are much smaller, by between one ($M_H \sim 50$ GeV) and almost three ($M_H \sim 1000$ GeV) orders of magnitude.

9.4 READYING FOR THE LHC

In view of the upcoming LHC, a list was compiled of the most promising signatures which should have allowed allow for Higgs detection at the LHC, as follows (wherein X represent additional particles):

- $gg \to H \to \gamma\gamma$,
- $q\bar{q}' \to WH \to \ell\nu_\ell\gamma\gamma$ and $gg, q\bar{q} \to t\bar{t}H \to \ell\nu_\ell\gamma\gamma X$,
- $q\bar{q}' \to WH \to \ell\nu_\ell b\bar{b}$ and $gg, q\bar{q} \to t\bar{t}H \to b\bar{b}b\bar{b}WW \to b\bar{b}b\bar{b}\ell\nu_\ell X$,
- $gg \to H \to Z^{(*)}Z^{(*)} \to \ell^+\ell^-\ell'^+\ell'^-$, where $\ell, \ell' = e$ or μ,
- $gg \to H \to W^{(*)}W^{(*)} \to \ell^+\nu_\ell\ell'^-\bar{\nu}_{\ell'}$, where $\ell, \ell' = e$ or μ,
- $gg \to H \to ZZ \to \ell^+\ell^-\nu_{\ell'}\bar{\nu}_{\ell'}$, where $\ell = e$ or μ and $\ell' = e, \mu$ or τ.

We are now ready to move on to tell the story of the Higgs boson discovery at the LHC, which we will do later, though. Before, in fact, we need to describe in some detail the fermion sector of the SM.

CHAPTER 10

Fermion Masses and Mixing

We have so far dealt with the construction of the gauge structure of the SM as well as introduced the Higgs boson mechanism for mass generation. It is now high time to start discussing its fermionic sector, namely, the matter side of this construction. In doing so, we will introduce the key concept of flavour mixing and from this we will move on to describe the GIM mechanism which led to the introduction of a second doublet of quarks, finally ending our discussion by introducing the need for CP violation.

10.1 LEPTON MASSES

We start our discussion of fermion masses by considering the lepton sector. In this sector, the Yukawa interactions are given by

$$\mathcal{L}_Y = Y_\ell \bar{\ell}_L \Phi \ell_R + h.c., \tag{10.1}$$

where $\ell_L = \begin{pmatrix} \nu \\ \ell \end{pmatrix}_L$ and $\Phi = \begin{pmatrix} \phi^+ \\ \phi^0 \end{pmatrix}$. As explained above, in the unitary gauge, the Higgs doublet takes the form $\Phi = \begin{pmatrix} 0 \\ \frac{v+H}{\sqrt{2}} \end{pmatrix}$, where H is the physical Higgs boson. Therefore, one finds

$$\mathcal{L}_Y = Y_\ell \bar{\ell}_L \Phi \ell_R + h.c. \tag{10.2}$$

DOI: 10.1201/9780429443015-10

In this case, the above Yukawa Lagrangian becomes

$$\mathcal{L}_Y = \frac{Y_\ell v}{\sqrt{2}}\bar{\ell}_L \ell_R + \frac{Y_\ell}{\sqrt{2}}\bar{\ell}_L H \ell_R + h.c. \tag{10.3}$$

The first term represents the charged lepton mass matrix, $M_\ell = Y_\ell v/\sqrt{2}$, while the second term is the Higgs interaction with leptons. It is remarkable that there is no mass term for the neutrinos nor interaction between neutrinos and the Higgs field.

If Y_ℓ is a 3×3 non-diagonal matrix, then one must perform the following unitary transformation to diagonalise M_ℓ:

$$\ell_L \to U\ell_L, \quad \ell_R \to V\ell_R, \tag{10.4}$$

such that

$$U^+ M_\ell V = M_\ell^{\rm diag} = \begin{pmatrix} m_e & & \\ & m_\mu & \\ & & m_\tau \end{pmatrix}. \tag{10.5}$$

Note that any matrix M can be diagonalised by a bi-unitary transformation, such that $M = U^+ m V$, where $m = \text{diag}\{m_1, m_2, ..., m_n\}$. This can be understood as follows. Since MM^+ is a Hermitian matrix, then it can be diagonalised with one unitary matrix U, *i.e.*, $MM^+ = U^+ m^2 U$. If we assume that M can be written as $M = U^+ m V$, then $V = m^{-1}UM$. Thus $VV^+ = (m^{-1}UM)(M^+U^+m^{-1}) = m^{-1}(UMM^+U^+)m^{-1} = m^{-1}m^2m^{-1} = 1$. So V is a unitary matrix and $M = U^+ m^{\rm diag} V$, *i.e.*, M is diagonalised by two unitary matrices, U and V.

10.2 QUARK MASSES

Now we consider quark masses. Let us start with down quarks. As mentioned above, the quark doublet q_L is defined as $q_L = \begin{pmatrix} u \\ d \end{pmatrix}_L$ and the quark singlets are $q_R \equiv \{u_R \text{ or } d_R\}$. The relevant Yukawa interaction is given by

$$\begin{aligned} \mathcal{L}_Y &= Y_d \bar{q}_L \Phi d_R + h.c. \\ &= Y_d (\bar{u}_L \ \bar{d}_L) \begin{pmatrix} 0 \\ \frac{v+H}{\sqrt{2}} \end{pmatrix} d_R + h.c. \\ &= \frac{Y_d v}{\sqrt{2}} \bar{d}_L d_R + \frac{Y_d}{\sqrt{2}} \bar{d}_L H d_R + h.c. \end{aligned} \tag{10.6}$$

Thus the down quark masses are given by $M_d = Y_d v/\sqrt{2}$. If Y_d is not diagonal, then we perform the following transformations (from gauge to mass eigenstates):

$$d_L \to V_L^d d_L, \quad d_R \to V_R^d d_R, \tag{10.7}$$

such that $V^{d^+} M_d V_R^d = M_d^{\text{diag}}$.

In order to give mass to the up quarks, as emphasised above, we need to introduce the conjugate Higgs doublet defined as

$$\tilde{\Phi} = i\tau_2 \Phi^* = \begin{pmatrix} \phi^{0^*} \\ -\phi^{+^*} \end{pmatrix}. \tag{10.8}$$

Note that $\tilde{\Phi}$ is also doublet of the $SU(2)$ group, since

$$\begin{aligned} \tilde{\Phi}' &= i\tau_2 \Phi'^* = i\tau_2 e^{-i\frac{\vec{\tau}^*}{2}.\vec{\Lambda}} \Phi^* = \left((i\tau_2) e^{-i\frac{\vec{\tau}^*}{2}.\vec{\Lambda}} (-i\tau_2)\right) (i\tau_2) \Phi^* \\ &= e^{-i\frac{\vec{\tau}^*}{2}.\vec{\Lambda}} \tilde{\Phi}. \end{aligned} \tag{10.9}$$

In the last step, we used the fact that $\tau_2(\vec{\tau}^*.\vec{\Lambda})\tau_2 = -\vec{\tau}.\vec{\Lambda}$. Since the hypercharge of Φ is 1, the hypercharge of $\tilde{\Phi}$ is -1. Now we can have the following Yukawa interactions:

$$\begin{aligned} \mathcal{L}_Y &= Y_u \bar{q}_L \tilde{\Phi} u_R + h.c. \\ &= Y_u (\bar{u}_L \ \bar{d}_L) \begin{pmatrix} \frac{v+H}{2} \\ 0 \end{pmatrix} u_R + h.c. \\ &= \frac{Y_u v}{\sqrt{2}} \bar{u}_L u_R + \frac{Y_u}{\sqrt{2}} \bar{u}_L H u_R + h.c. \end{aligned} \tag{10.10}$$

Thus the up-quark masses are given by $M_u = Y_u v/\sqrt{2}$. If Y_u is not diagonal, then we perform the transformations

$$u_L \to V_L^u u_L, \quad u_R \to V_R^u u_R, \tag{10.11}$$

such that $V^{u^+} M_u V_R^u = M_u^{\text{diag}}$.

Finally, it is remarkable that Higgs boson couplings to quarks, in mass eigenstates, are given by

$$\mathcal{L}_{Hqq} = \frac{-H}{v}\Big(m_u \bar{u}u + m_d \bar{d}d + m_s \bar{s}s + m_c \bar{c}c + m_b \bar{b}b + m_t \bar{t}t\Big), \tag{10.12}$$

which are also flavour diagonal.

10.3 FLAVOUR MIXING

The quark charged current is given by the following Lagrangian:

$$\begin{aligned} \mathcal{L}_{\text{CC}} &= -\frac{g}{\sqrt{2}}(\bar{u}_L \bar{d}_L)_i \gamma^\mu \tau^+ W_\mu^+ \begin{pmatrix} u_L \\ d_L \end{pmatrix}_i \\ &= -\frac{g}{\sqrt{2}}(\bar{u}_L)_i \gamma^\mu W_\mu^+ (d_L)_i + h.c. \end{aligned} \tag{10.13}$$

After the above mentioned rotations that diagonalise up- and down-quark mass matrices, the quark charged current (see Fig. 10.1) takes the form

$$\begin{aligned} \mathcal{L}_{\text{CC}} &= -\frac{g}{\sqrt{2}}(\bar{u'}_L)_\alpha [(V_L^{u^+})_{\alpha i}(V_L^{d^+})_{i\alpha}] \gamma^\mu W_\mu^+ (d_L)_\beta + h.c. \\ &= -\frac{g}{\sqrt{2}}(\bar{u'}_L)_\alpha V_{\alpha\beta}^{\text{CKM}} \gamma^\mu W_\mu^+ (d_L)_\beta + h.c., \end{aligned} \tag{10.14}$$

where we have defined the mixing matrix, V^{CKM}, as

$$V^{\text{CKM}} = V_L^{u^+} V_L^d. \tag{10.15}$$

This is not a diagonal matrix, due to the mismatch between up- and down-quark rotations. However it is unitary. In general, V^{CKM} has 9 real parameters which can be reduced to 4: three "rotational angles" and one phase. This is the famous CKM matrix[1] and hence the charged current is flavour dependent.

[1]Which was worth the 2008 Nobel Prize in Physics to Makoto Kobayashi and Toshihide Maskawa "for the discovery of the origin of the broken symmetry which predicts the existence of at least three families of quarks in nature" [118], shared with Yoichiro

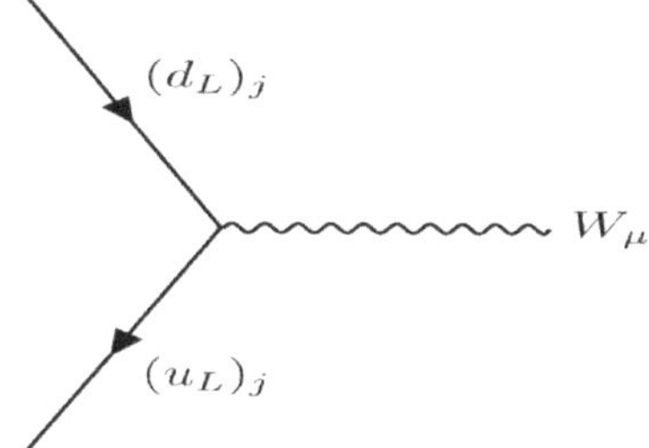

Figure 10.1 Quark charged current.

In case of two generations, as we will show in the next chapter, $V^{\rm CKM}$ has only one real parameter, which is the Cabibbo angle, and there is no freedom to have a CP violating phase in the ensuing mixing matrix.

The lepton charged current is given by

$$\begin{aligned} \mathcal{L}_{\rm CC} &= -\frac{g}{\sqrt{2}}(\bar{\nu}_L \bar{\ell}_L)_i \gamma^\mu \tau^+ W_\mu^+ \begin{pmatrix} \nu_L \\ \ell_L \end{pmatrix}_i \\ &= -\frac{g}{\sqrt{2}}(\bar{\nu}_L)_i \gamma^\mu W_\mu^+ (\ell_L)_i + h.c. \end{aligned} \tag{10.16}$$

In mass eigenstates, the above Lagrangian transforms to

$$\mathcal{L}_{\rm CC} = -\frac{g}{\sqrt{2}}(\bar{\nu}_L)_i \gamma^\mu W_\mu^+ (U\ell_L)_i + h.c. \tag{10.17}$$

Since neutrinos are massless there is no restriction on the redefinition of the neutrino fields and we can define $\nu_L' = U\nu_L$. Hence, we will not have any mismatch between the rotation of charged leptons and neutrinos. Thus, the charged current in the lepton sector is flavour diagonal: see Fig. 10.2.

Concerning the neutral current, it remains flavour diagonal as can be seen from the following Lagrangian:

$$\begin{aligned} \mathcal{L}_{\rm NC} &= e\left(\bar{u}\gamma_\mu u + \bar{d}\gamma_\mu d\right) A^\mu + \frac{g}{\cos\theta_W}\Big[g_L^u \bar{u}_L \gamma_\mu u_L \\ &+ g_L^d \bar{d}_L \gamma_\mu d_L + g_R^u \bar{u}_R \gamma_\mu u_R + g_R^d \bar{d}_R \gamma_\mu d_R\Big], \end{aligned} \tag{10.18}$$

Nambu "for the discovery of the mechanism of spontaneous broken symmetry in subatomic physics" [119].

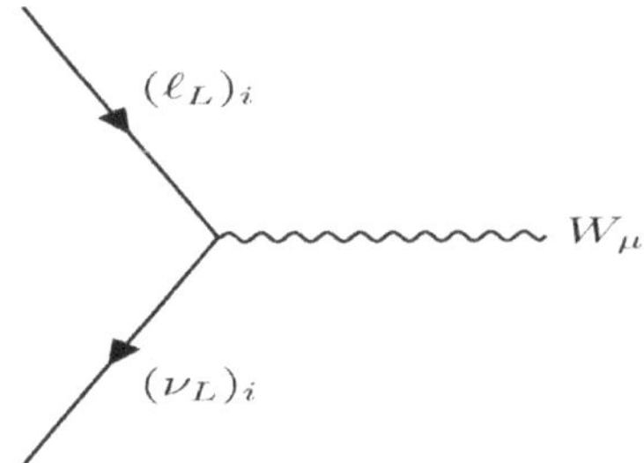

Figure 10.2 Lepton charged current.

where $g_R^u = 4/3$ and $g_R^d = -2/3$. So there is no mismatch rotation in any term and the quark neutral current in the SM is flavour diagonal: see Fig. 10.3.

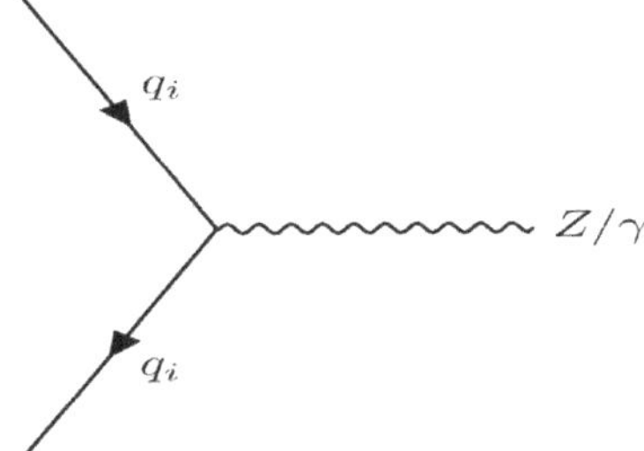

Figure 10.3 Quark neutral current.

The neutral current in the lepton sector is described by the following Lagrangian:

$$\mathcal{L}_{\rm NC} = e\left(\bar{\ell}\gamma_\mu \ell\right) A^\mu + g\left[g_L^\nu \bar{\nu}_L \gamma_\mu \nu_L + g_L^\ell \bar{\ell}_L \gamma_\mu \ell_L + g_R^\ell \bar{\ell}_R \gamma_\mu \ell_R\right], \qquad (10.19)$$

where $g_L^\nu = 1/2$, $g_L^\ell = -1/2 + \sin^2\theta_W$ and $g_R^\ell = \sin^2\theta_W$. Recall that $g_R^\nu = 0$. Since the rotation $\ell \to U\ell$ does not change $\mathcal{L}_{\rm NC}$, the neutral currents of leptons is also flavour diagonal (see Fig. 10.4).

10.4 CKM MIXING MATRIX

An $n \times n$ unitary matrix V is characterised by n^2 real parameters, as the unitarity condition imposes n^2 relations among the $2n^2$ parameters of V. Then V has $n(n-1)/2$ angles and $n^2 - n(n-1)/2 = n(n+1)/2$ phases. Some of these

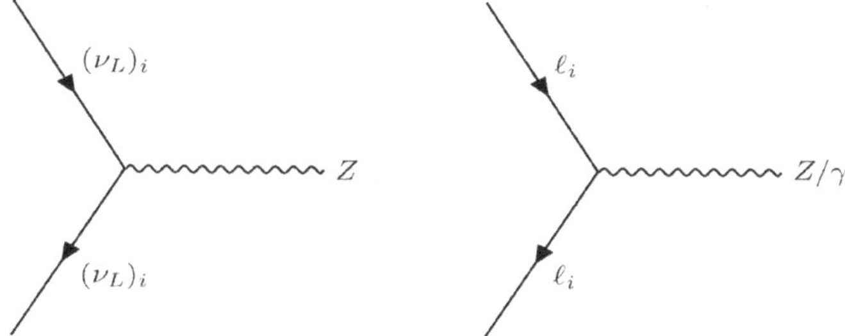

Figure 10.4 Lepton neutral current.

phases can be absorbed into new quark fields. Indeed, $2n-1$ of these phases can be removed by quark rephasing: $u_{L_\alpha} \to e^{i\theta^u_\alpha} u_{L_\alpha}$ and $d_{L_\alpha} \to e^{i\theta^d_\alpha} d_{L_\alpha}$. Thus $V_{\alpha\beta} \to V_{\alpha\beta} e^{i(\theta^d_\beta - \theta^u_\alpha)}$, for $\alpha, \beta = 1, ..., n$.

The CKM matrix is a unitary matrix, since $V_{\rm CKM} V^\dagger_{\rm CKM} = (V^\dagger_u V_d)(V^\dagger_d V_u) = 1$. Thus, for three generations of quarks, the CKM matrix generally depends on 9 parameters: 3 real angles and 6 phases. However, 5 of these phases can be absorbed into the redefinition of the quark fields. Therefore, in the SM with three families, CP violation is accommodated by a single complex phase in the CKM mixing matrix. In contrast, one can conclude that, with two quark generations, no CP violation can be generated. Indeed, a 2×2 unitary matrix is a real matrix, where all phases are rotated away by some field redefinitions. For example, the first column, $V_{\alpha 1]}$ $\alpha = 1, ..., n$, can be made real by redefining the up-type quark fields as $u'_{L_\alpha} = e^{i\theta^u_\alpha} u_{L_\alpha}$ and the first row $V_{1\beta}$, $\beta = 2, 3, ..., n$ (as already V_{11} is made real) can be made real by $d'_{L_\beta} = e^{i\theta^d_\beta} d_{L_\beta}$. So $(2n-1)$ phases are removed from V without changing physical consequences. Thus, the remaining parameters are characterised by $n(n-1)/2$ angles and $n(n+1/2 - (2n-1) = (n-1)(n-2)/2$ phases.

In case of $n = 2$, *i.e.*, four quarks only, the mixing matrix V can be parametrised by only one flavour mixing angle θ_C, called Cabibbo angle. There is no phase since $(n-1)(n-2)/2 = 0$ for $n = 2$. Hence, we have no CP violation. The charged current can be written as

$$\mathcal{L} = \frac{g}{\sqrt{2}} (\bar{u}\ \bar{c}) \gamma^\mu \frac{(1-\gamma_5)}{2} V \begin{pmatrix} d \\ s \end{pmatrix} W^+_\mu + h.c., \tag{10.20}$$

with

$$V = \begin{pmatrix} V_{ud} & V_{us} \\ V_{cd} & V_{cs} \end{pmatrix} = \begin{pmatrix} \cos\theta_C & \sin\theta_C \\ -\sin\theta_C & \cos\theta_C \end{pmatrix}.$$

This means that the down quark mass state is given by

$$\begin{pmatrix} d_c \\ s_c \end{pmatrix}_L = \begin{pmatrix} \cos\theta_C & \sin\theta_C \\ -\sin\theta_C & \cos\theta_C \end{pmatrix} \begin{pmatrix} d \\ s \end{pmatrix}_L \qquad (10.21)$$

and the weak charged current processes are due to the transition $u \leftrightarrow d_c$ and $c \leftrightarrow s_c$.

In case of $n = 3$ (six quarks), the mixing matrix $V_{\rm CKM}$ will have 3 angles and 1 phase. Thus, we expect CP violation, accordingly. There are several parameterisations of the CKM matrix which are physically equivalent. The standard parameterisation of $V_{\rm CKM}$ is given by [120]

$$V_{\rm CKM} = \begin{pmatrix} c_{12}c_{13} & s_{12}c_{13} & s_{13}e^{-i\delta} \\ -s_{12}c_{23} - c_{12}s_{23}s_{13}e^{i\delta} & c_{12}c_{23} - s_{12}s_{23}s_{13}e^{i\delta} & s_{23}c_{13} \\ s_{12}s_{23} - c_{12}23s_{13}e^{i\delta} & -c_{12}s_{23} - s_{12}c_{23}s_{13}e^{i\delta} & c_{23}c_{13} \end{pmatrix}, \qquad (10.22)$$

where $c_{ij} = \cos\theta_{ij}$, $s_{ij} = \sin\theta_{ij}$ and δ is the CP violating phase. Another useful parameterisation of the CKM matrix was introduced by Wolfenstein [121] and is based on experimental results and unitarity to express $V_{\rm CKM}$ as a series expansion in $\lambda \equiv \sin\theta_C \simeq 0.22$:

$$V_{\rm CKM} = \begin{pmatrix} 1-\frac{1}{2}\lambda^2 & \lambda & A\lambda^3(\rho - i\eta) \\ -\lambda & 1-\frac{1}{2}\lambda^2 & A\lambda^2 \\ A\lambda^3(1-\rho-i\eta) & -A\lambda^2 & 1 \end{pmatrix} + \mathcal{O}(\lambda^4). \qquad (10.23)$$

The parameter A is determined by measuring V_{cb} from the inclusive leptonic b decays and the exclusive decay $B^0 \to D^{*+}l^-\bar{\nu}_l$. The (ρ, η) plane can be constrained from the combination of many experimental observables as in Fig. 10.5. Note that $\bar{\rho} = \rho(1-\lambda^2/2)$ and $\bar{\eta} = \eta(1-\lambda^2/2)$ are usually used instead of ρ and η to keep unitarity exact. The best determination of the Wolfenstein parameters is [122]

$$\begin{aligned} \rho &= 0.135^{+0.031}_{-0.016}, \qquad \eta = 0.349^{+0.015}_{-0.017}, \qquad A = 0.814^{+0.021}_{-0.022}, \\ \lambda &= 0.2257^{+0.0009}_{-0.0010}. \end{aligned} \qquad (10.24)$$

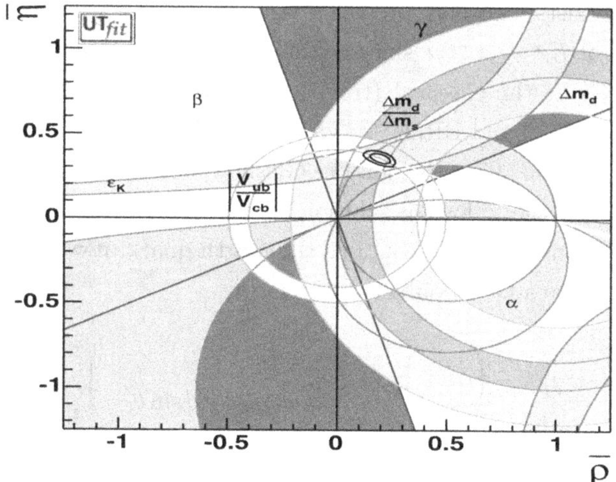

Figure 10.5 Experimental constraints on the (ρ, η) plane as obtained by the UTfit Collaboration [122].

The uncertainty is still not negligible, in particular, the theoretical results suffer from significant uncertainties due to the hadronic matrix elements involved in the semileptonic decays, as we will show in the next chapter.

10.5 GIM MECHANISM

In 1960s, only one quark doublet was known from the analysis of β-decay, which occurred via the weak charged current transition $u \leftrightarrow d_c$, where d_c is a down-quark in mass eigenstate, between the members of the doublet

$$Q_L^{(1)} = \begin{pmatrix} u \\ d_c \end{pmatrix}_L = \begin{pmatrix} u \\ d\cos\theta_C + s\sin\theta_C \end{pmatrix}_L.$$

At that time, the orthogonal combination of d_L and s_L, *i.e.*, $s_c = s\cos\theta_C - d\sin\theta_C$ was left as $SU(2)_L$ singlet. Then, when we apply this to neutral currents, we have

$$\begin{aligned} \bar{Q^{(1)}}_L \gamma_\mu \frac{\tau^3}{2} Q_L^{(1)} &= \frac{1}{2}\Big[\bar{u}_L\gamma_\mu u_L - \cos^2\theta_C \bar{d}_L\gamma_\mu d_L - \sin^2\theta_C \bar{s}_L\gamma_\mu s_L \\ &- \cos\theta_C\sin\theta_C(\bar{d}_L\gamma_\mu s_L + \bar{s}_L\gamma_\mu d_L)\Big], \end{aligned} \tag{10.25}$$

where the final term corresponds to a strangeness changing neutral current. If this really exists, one may predict a large decay width of FCNCs such as $K_L^0 \to \mu^+\mu^-$ or $K^+ \to \pi^+\bar{\nu}\nu$, since the flavour-changing $\bar{s}dZ$ and $\bar{d}sZ$ vertices appear already at the classical (tree) level. However, $\mathrm{BR}(K_L^0 \to \mu^+\mu^-) = (7.25 \pm 0.16) \times 10^{-9}$ and $\mathrm{BR}(K^+\pi^+\bar{\nu}\bar{\nu} = (1.6^{+1.8}_{-0.8}) \times 10^{-10}$.

As mentioned, in 1970, Sheldon Lee Glashow, John Iliopoulos and Luciano Maiani proposed the following solution to this problem. They introduced a second quark doublet, which contains the fourth quark, now called charm, c, quark. So we have a $Q_L^{(2)}$ doublet in the form:

$$Q_L^{(2)} = \begin{pmatrix} c \\ s_c \end{pmatrix}_L = \begin{pmatrix} c \\ s\cos\theta_C - d\sin\theta_C \end{pmatrix}_L ,$$

which produces additional neutral current:

$$\begin{aligned} \bar{Q^{(2)}}_L \gamma_\mu \frac{\tau^3}{2} Q_L^{(2)} &= \frac{1}{2}\Big[\bar{c}_L\gamma_\mu c_L - \cos^2\theta_C \bar{s}_L\gamma_\mu s_L + \sin^2\theta_C \bar{d}_L\gamma_\mu d_L \\ &+ \cos\theta_C\sin\theta_C(\bar{d}_L\gamma_\mu s_L + \bar{s}_L\gamma_\mu d_L)\Big]. \end{aligned} \tag{10.26}$$

Summing up $\bar{Q^{(1)}}_L\gamma_\mu\frac{\tau^3}{2}Q_L^{(1)}$ and $\bar{Q^{(2)}}_L\gamma_\mu\frac{\tau^3}{2}Q_L^{(2)}$, one finds that the FCNC becomes flavour diagonal, given by

$$\frac{1}{2}\Big[\bar{u}_L\gamma_\mu u_L + \bar{c}_L\gamma_\mu c_L - \bar{d}_L\gamma_\mu d_L - \bar{s}_L\gamma_\mu s_L\Big]$$

and we have no FCNCs at the tree level.

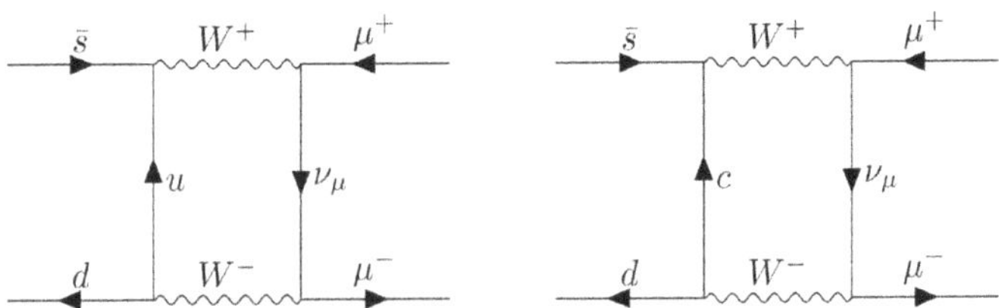

Figure 10.6 GIM mechanism and $K_L^0 \to \mu^+\mu^-$ process.

Then they proposed that FCNCs will occur at the quantum level via the charged current. For example $K_L^0 \to \mu^+\mu^-$ can take place via the Feynman diagrams in Fig. 10.6. The amplitude of the first diagram is given by

$$A(K_L^0 \mu^+ \mu^-) \propto \frac{g^4 \sin\theta_C \cos\theta_C}{M_W^2} \simeq \alpha G_F \sin\theta_C \cos\theta_C, \tag{10.27}$$

where $\alpha = e^2/4\pi$. The second diagram amplitude is proportional to $-g^4 \sin\theta_C \cos\theta_C \,/M_W^2$, which is just opposite to the first contribution. Therefore, the amplitude exactly vanishes at the leading order of perturbation in m_q^2/M_W^2. The non-vanishing contribution stems from the first order of m_q^2/M_W^2. It turns out to be

$$A(K_L^0 \to \mu^+ \mu^-) \simeq \frac{g^2 \sin\theta_C \cos\theta_C}{M_W^2} \times \frac{m_c^2 - m_u^2}{M_W^2}. \tag{10.28}$$

Thus for $\frac{m_c^2 - m_u^2}{M_W^2} \sim 3 \times 10^{-4}$, one gets the BR$(K_L^0 \mu^+ \mu^-)$ of the right order of magnitude. This leads to $m_c \simeq 1.5$ GeV, which was a very good prediction, matched by the experimental results found later.

CHAPTER 11

CP Violation

Symmetries and conservation laws play a crucial role in particle physics. There are two type of symmetries: continuous and discrete. The invariance of a system under a continuous symmetry transformation leads to a conservation law, this is known as the Noether theorem. For example, the invariance under space and time translations results in momentum and energy conservation, respectively. The gauge symmetries, $SU(3)_C$, $SU(2)_L$ and $U(1)_Y$, considered in the previous chapters are examples of continuous internal symmetries. Here we focus on three important examples of discrete symmetries that are a topic of interest in modern particle physics: C, P and T[1].

As mentioned above, parity transformation P performs a reflection of the space coordinates at the origin:

$$P\psi(\vec{r}) = \psi(-\vec{r}).$$

Applying parity twice restores the original state, $P^2 = 1$. Therefore, the parity of a wavefunction $\psi(\vec{r})$ has to be either even, $P = +1$, or odd, $P = -1$. The invariance of the Dirac equation under parity transformation implies that

$$\psi(\vec{r}, t) = P\psi(-\vec{r}, t) = \gamma^0\psi(-\vec{r}, t).$$

P is a multiplicative quantum number, so the parity of a many particle system is equal to the product of the intrinsic parities of the particles times the parity of the spatial wavefunction which is $(-1)^L$.

[1] We will use the italics C, P and T for the corresponding operators.

DOI: 10.1201/9780429443015-11

C is another discrete symmetry that reverses the sign of the electric charge, colour charge and magnetic moment of a particle. It also reverses the values of the weak isospin and hypercharge charges associated with the weak force. Like the parity operator it satisfies $C^2 = 1$ and has possible eigenvalues $C = \pm 1$. The invariance of the Dirac equation under charge conjugation implies that

$$\psi(\vec{r}, t) \longrightarrow \psi_c(\vec{r}, t) = i\gamma^2 \psi^*(\vec{r}, t) = C\bar{\psi}^T(\vec{r}, t),$$

with $C = i\gamma^2\gamma^0$. For fermions charge conjugation changes a particle into an antiparticle, so fermions themselves are not eigenstates of C. Combinations of fermions can be eigenstates of C. For example, positronium has $C(e^+e^-) = (-1)^{L+S}$, where S is the sum of the spins which can be either 0 or 1.

Finally, T is another discrete symmetry, defined as $T\psi(t) = \psi(-t)$ with $T^2 = 1$, and possible eigenvalues are again $T = \pm 1$. The solutions of the Dirac equation describe antifermion states as equivalent to fermion states with the time and space coordinates reversed. The invariance of the Dirac equation under time reversal implies that

$$\psi(\vec{r}, t) \longrightarrow \psi_{_T}(\vec{r}, t) = i\gamma^1\gamma^3 \psi^*(\vec{r}, -t) = T\psi^*(\vec{r}, -t).$$

The combination of C, P and T, known as CPT, is an exact symmetry in QFT, due to the assumptions of locality, Lorentz invariance and hermiticity of the Lagrangian, *i.e.*,

$$(CPT)\ \mathcal{L}(x, t)\ (CPT)^{-1} = \mathcal{L}(-x, -t). \tag{11.1}$$

11.1 CP VIOLATION AND CKM MATRIX

It turns out that both C and P are maximally violated in weak decays. However, experimental results suggest that the combination CP is a nearly conserved symmetry. CP turns a particle into its antiparticle with opposite helicity: it is a symmetry between matter and antimatter. It is a conserved quantity in strong and EM interactions. Finally, CP violation requires the presence of a complex phase.

The SM Lagrangian consists of three parts. The first one, which is known as the gauge sector, is the part of the Lagrangian that contains the interactions of fermions with the gauge bosons and also the interactions of the gauge bosons among themselves. The second part is the Yukawa sector, which contains the interactions of fermions with the scalar Higgs doublet. The third part is the Higgs sector, which contains the scalar potential of the Higgs field. It is remarkable that the gauge sector is CP conserving and hence there is no complex phase in this sector. Also the Higgs sector of one Higgs doublet models can be made real by field redefinitions. Therefore, there is no CP violating phase in this sector either. Finally, the Yukawa couplings are generally complex and, as we will show, they are indeed the source of CP violation in the SM.

The unitarity of the CKM matrix leads to the following relations:

$$V_{ud}V_{us}^* + V_{cd}V_{cs}^* + V_{td}V_{ts}^* = 0, \tag{11.2}$$

$$V_{us}V_{ub}^* + V_{cs}V_{cb}^* + V_{ts}V_{tb}^* = 0, \tag{11.3}$$

$$V_{ud}V_{ub}^* + V_{cd}V_{cb}^* + V_{td}V_{tb}^* = 0. \tag{11.4}$$

Each of these relations can be represented graphically by a triangle. The last relation is commonly used to define what is called the 'unitarity triangle'. One can display this triangle in terms of the following three angles:

$$\alpha = \arg\left(-\frac{V_{td}V_{tb}^*}{V_{ud}V_{ub}^*}\right), \quad \beta = \arg\left(-\frac{V_{cd}V_{cb}^*}{V_{td}V_{tb}^*}\right), \quad \gamma = \arg\left(-\frac{V_{ud}V_{ub}^*}{V_{cd}V_{cb}^*}\right), \tag{11.5}$$

which satisfy the constraint $\alpha+\beta+\gamma=\pi$, as shown in Fig. 10.5. Multiplying Eq. (11.4) by $V_{us}^*V_{cs}$ and taking the imaginary part, one finds that

$$\mathrm{Im}\,[V_{us}^*V_{cs}V_{ud}V_{cd}^*] = -\mathrm{Im}\,[V_{us}^*V_{cs}V_{ub}V_{cb}^*]\,. \tag{11.6}$$

Also, from other orthogonal relations, one can show that the quartets $V_{ij}V_{kl}V_{il}^*V_{kj}^*$ have equal imaginary parts up to a sign. Therefore, one defines the Jarlskog invariant [123, 124]

$$J = \mathrm{Im}\,[V_{us}V_{cb}V_{ub}^*V_{cs}^*] \tag{11.7}$$

as an invariant measure for the amount of CP violation in the SM. Using the Wolfenstein parameterisation for V_{CKM}, one can show that $J \simeq A^2\lambda^6\eta \lesssim 10^{-4}$. It is remarkable that the smallness of J is due to the smallness of the off-diagonal elements of V_{CKM}, in particular $|V_{ub}|$, and not because of a small value of the CP violating phase. In addition, the area of the unitarity triangle is given by $|J|/2$. Thus, the area of this triangle measures the amount of CP violation.

In quark mass eigenstates, the charged current interactions are given by

$$\begin{aligned} -\mathcal{L}_{W^+} &= \frac{g}{\sqrt{2}} \bar{u}_{L_i} \gamma^\mu (V_{uL} V_{dL}^\dagger)_{ij} d_{Lj} W_\mu^+ + h.c. \\ &= \frac{g}{\sqrt{2}} \bar{u}_{L_i} \gamma^\mu (V_{\text{CKM}})_{ij} d_{Lj} W_\mu^+ + h.c. \end{aligned} \tag{11.8}$$

Hence, the quark generations (flavours) are mixed within the charged interactions, while in the neutral current interactions there is no such a mixing and flavour remains universal. This can be understood from the following:

$$\begin{aligned} -\mathcal{L}_Z = &\frac{g}{\cos\theta_W} \left[\Big(\frac{1}{2} - \frac{2}{3}\Big) \bar{u}_{L_i} \gamma^\mu (V_{uL} V_{uL}^\dagger)_{ij} u_{Lj} \right. \\ &\left. + \Big(\frac{-1}{2} + \frac{2}{3}\Big) \bar{d}_{L_i} \gamma^\mu (V_{dL} V_{dL}^\dagger)_{ij} d_{Lj} \right] Z_\mu \\ = &\frac{g}{\cos\theta_W} \left[\Big(\frac{1}{2} - \frac{2}{3}\Big) \bar{u}_{L_i} \gamma^\mu u_{Li} + \Big(\frac{-1}{2} + \frac{2}{3}\Big) \bar{d}_{L_i} \gamma^\mu d_{Li} \right] Z_\mu . \end{aligned} \tag{11.9}$$

Thus, one may conclude that CP is explicitly broken in the SM by the phase δ in the CKM mixing matrix. Also, both flavour and CP violation appear only in the charged current interactions of quarks.

11.2 CP VIOLATION IN NEUTRAL MESONS

In this section we consider the mixing and decays of P^0 and $\bar{P^0}$. These mesons states are flavour eigenstates, with opposite flavour quantum numbers. The CP transformation interchanges P^0 and $\bar{P}^0$, such that $(CP)^2 = 1$, *i.e.*,

$$CP|P^0\rangle = e^{i\zeta}|\bar{P}^0\rangle, \tag{11.10}$$

$$CP|\bar{P}^0\rangle = e^{-i\zeta}|P^0\rangle. \tag{11.11}$$

It is clear that $|P^0\rangle$ and $|\bar{P^0}\rangle$ are not CP eigenstates. However, the corresponding CP eigenstates can be constructed as follows:

$$|P_1\rangle = \frac{e^{-i\zeta}|P^0\rangle + |\bar{P}^0\rangle}{\sqrt{2}}, \tag{11.12}$$

$$|P_2\rangle = \frac{e^{-i\zeta}|P^0\rangle - |\bar{P}^0\rangle}{\sqrt{2}}. \tag{11.13}$$

Now, $|P_1\rangle$ and $|P_2\rangle$ have CP eigenvalues $+1$ and -1, respectively. Moreover, $|P^0\rangle$ and $|\bar{P}^0\rangle$ are not mass eigenstates. A superposition state of P^0 and $\bar{P}^0$, given by

$$\psi(t) = a(t)|P^0\rangle + b(t)|\bar{P}^0\rangle, \tag{11.14}$$

evolves according to the Schrodinger equation:

$$i\frac{d}{dt}\begin{pmatrix} a \\ b \end{pmatrix} = H\begin{pmatrix} a \\ b \end{pmatrix}. \tag{11.15}$$

The Hamiltonian is in general not Hermitian for unstable particles. It can be written as

$$H = M - \frac{i}{2}\Gamma,$$

where M and Γ are 2×2 Hermitian matrices. The off-diagonal elements M_{12} and $\frac{i}{2}\Gamma_{12}$ describe the transition $P^0 \to \bar{P}^0$. Also due to CPT conservation, one finds $M_{11} = M_{22}$, $M_{21} = M_{12}^*$ and $\Gamma_{11} = \Gamma_{22}$, $\Gamma_{21} = \Gamma_{12}^*$, which imply that particle and antiparticle have equal mass and total decay width.

The eigenvalues of this Hamiltonian, $\lambda_{L,H}$ are given by

$$\lambda_{L,H} = M - \frac{i}{2}\Gamma \pm \sqrt{\left(M_{12}^* - \frac{i}{2}\Gamma_{12}^*\right)\left(M_{12} - \frac{i}{2}\Gamma_{12}\right)}. \tag{11.16}$$

The corresponding eigenstates, light P_L and heavy P_H mesons, can be defined as

$$|P_{L,H}\rangle = p|P^0\rangle \pm q|\bar{P}^0\rangle,$$

where p and q are obtained by solving the following equation:

$$\begin{pmatrix} M - \frac{i}{2}\Gamma & M_{12} - \frac{i}{2}\Gamma_{12} \\ M_{12}^* - \frac{i}{2}\Gamma_{12}^* & M - \frac{i}{2}\Gamma \end{pmatrix}\begin{pmatrix} p \\ q \end{pmatrix} = \lambda_{L,H}\begin{pmatrix} p \\ q \end{pmatrix}. \tag{11.17}$$

One can easily find that q/p is given by

$$\frac{q}{p} = \pm\sqrt{\frac{M_{12}^* - \frac{i}{2}\Gamma_{12}^*}{M_{12} - \frac{i}{2}\Gamma_{12}}}. \tag{11.18}$$

Also, the mass difference $\Delta M = M_H - M_L$ is given by

$$\Delta M = 2\,\mathrm{Re}\sqrt{\left(M_{12}^* - \frac{i}{2}\Gamma_{12}^*\right)\left(M_{12} - \frac{i}{2}\Gamma_{12}\right)} \simeq 2|M_{12}|. \tag{11.19}$$

In the last equality, $\Gamma_{12} \ll M_{12}$ was assumed. Similarly, the width difference $\Delta\Gamma = \Gamma_H - \Gamma_L$ is given by

$$\Delta\Gamma = -2\frac{\mathrm{Re}\left[M_{12}\Gamma_{12}^*\right]}{|M_{12}|}. \tag{11.20}$$

At the LO in $|\Gamma_{12}/M_{12}|$, the complex parameter q/p, which is relevant for analysing the CP violation in mixing, can be written as

$$\frac{q}{p} = \frac{M_{12}^*}{|M_{12}|}\left[1 - \frac{1}{2}\mathrm{Im}\left(\frac{\Gamma_{12}}{M_{12}}\right)\right]. \tag{11.21}$$

The mass eigenstates can be represented in terms of CP eigenstates as follows:

$$|P_{L,H}\rangle = \frac{e^{i\zeta}p \pm q}{\sqrt{2}}|P_1\rangle + \frac{e^{i\zeta}p \mp q}{\sqrt{2}}|P_2\rangle. \tag{11.22}$$

Therefore, CP is conserved (*i.e.*, the CP operator commutes with the Hamiltonian H) if and only if

$$e^{i\zeta} = \frac{q}{p}, \;\; i.e., \left|\frac{q}{p}\right| = 1 \implies \mathrm{Im}\left(\frac{\Gamma_{12}}{M_{12}}\right) = 0. \tag{11.23}$$

11.3 TYPES OF CP VIOLATION

As we will see, there are three types of CP violation in meson systems.

1. Indirect CP violation (CP violation in mixing): this type of CP violation occurs when $|q/p| \neq 1$, *i.e.*, the relative phase between M_{12} and Γ_{12} is non-zero. In this case the mass eigenstates are mixtures of CP eigenstates. Thus, the rate of $P^0 \to \bar{P}^0$ is not equal to the rate of $\bar{P}^0 \to P^0$.

2. Direct CP violation (CP violation in decay): this type of CP violation in mixing occurs when the amplitudes of P^0 and $\bar{P}^0$ decays have different magnitudes. If we define the decay amplitudes as

$$A_f = \langle f|\mathcal{H}|P^0\rangle, \;\; \bar{A}_f = \langle f|\mathcal{H}|\bar{P}^0\rangle, \tag{11.24}$$

where $\mathcal{H}$ is the weak Hamiltonian and f is the final state. Also we may have the amplitudes

$$A_{\bar{f}} = \langle \bar{f}|\mathcal{H}|P^0\rangle, \;\; \bar{A}_{\bar{f}} = \langle \bar{f}|\mathcal{H}|\bar{P}^0\rangle. \tag{11.25}$$

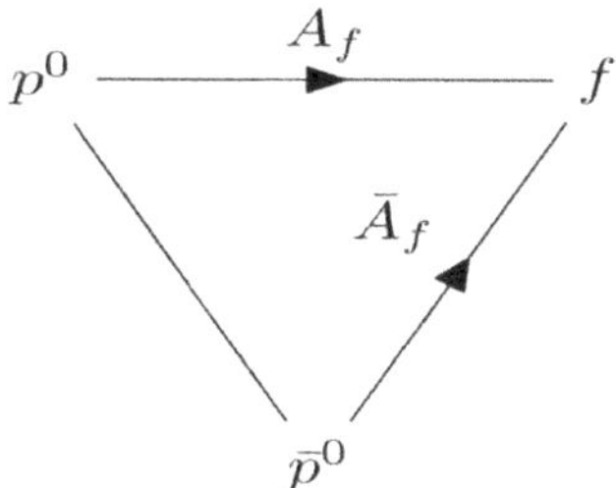

Figure 11.1 The decay of P^0 to final state f with/without mixing with $\bar{P}^0$.

If

$$CP|f\rangle = e^{i\zeta_f}|\bar{f}\rangle, \quad CP|\bar{f}\rangle = e^{-i\zeta_f}|f\rangle,$$

then the relative phase between A_f and $\bar{A}_{\bar{f}}$ is given by

$$\bar{A}_{\bar{f}} = e^{i(\zeta_f - \zeta)} A_f.$$

CP would be conserved in these decay process if and only if A_f and $\bar{A}_{\bar{f}}$ have the same magnitude. Thus $|\bar{A}_{\bar{f}}/A_f| \neq 1$ is an indication for CP violation in the meson decay. This process could include a CP conserving strong phase, δ, in addition to the CP violating weak phase(s). If the amplitude of $P^0 \to f$ as $\sum_i |A_i| e^{i(\delta_i + \phi_i)}$ and the amplitude of $\bar{P}^0 \to \bar{f}$ as $\sum_i |A_i| e^{i(\delta_i - \phi_i)}$ (note the changing of the sign of the weak phase in the CP conjugate process), then

$$\left| \frac{\bar{A}_{\bar{f}}}{A_f} \right| = \left| \frac{\sum_i |A_i| e^{i(\delta_i - \phi_i)}}{\sum_j |A_j| e^{i(\delta_j + \phi_j)}} \right|. \tag{11.26}$$

It is clear that the interference between at least two amplitudes with different weak phases would imply that $|\bar{A}_{\bar{f}}/A_f| \neq 1$.

3. CP violation in interference of mixing and decay: this type of CP violation takes place in processes where both P^0 and $\bar{P}^0$ decay to a final state f, as shown in Fig. 11.1. We introduce the complex parameter λ_f, defined as

$$\lambda_f = \frac{q}{p} \frac{\bar{A}_f}{A_f}. \tag{11.27}$$

Any value of $\lambda_f \neq 1$ is a manifestation of CP violation. Note that $\mathrm{Im}(\lambda_f) \neq 0$ even if direct or indirect CP violation vanishes.

11.4 CP VIOLATION IN THE KAON SYSTEM

In the K^0 and $\bar{K}^0$ system, the flavour eigenstates are given by $K^0 = (\bar{s}d)$ and $\bar{K}^0 = (s\bar{d})$. These states are not definite CP eigenstates but $\mathrm{CP}|K^0\rangle = -|\bar{K}^0\rangle$. A definite CP eigenstate is constructed as

$$K_{1,2} = \frac{1}{\sqrt{2}}\left(K^0 \mp \bar{K}^0\right) \tag{11.28}$$

with $\mathrm{CP}|K_{1,2}\rangle = \pm|K_{1,2}\rangle$. Therefore, we have $K_1 \to \pi\pi$ and $K_2 \to \pi\pi\pi$. As discussed above, the physical eigenstates of the Hamiltonian can be defined as

$$K_S = pK^0 + q\bar{k}^0 \tag{11.29}$$

and

$$K_L = pK^0 - q\bar{k}^0, \tag{11.30}$$

where the indices L and S refer to long (corresponds to heavy in the above definition) and short (corresponds to light in the above definition) mass eigenstates, respectively, and

$$p = (1+\varepsilon)/\sqrt{2(1+|\varepsilon|^2)}, \qquad q = (1-\varepsilon)/\sqrt{2(1+|\varepsilon|^2)}, \tag{11.31}$$

where ε is the CP violating parameter in the $K^0 - \bar{K}^0$ system, since $K_{S,L}$ can now be written as

$$K_S = \frac{1}{\sqrt{1+|\varepsilon|^2}}\left(K_1 + \varepsilon K_2\right), \tag{11.32}$$

$$K_L = \frac{1}{\sqrt{1+|\varepsilon|^2}}\left(K_2 + \varepsilon K_1\right). \tag{11.33}$$

Therefore, one finds

$$\frac{q}{p} = \frac{1-\varepsilon}{1+\varepsilon} = \sqrt{\frac{M_{12}^* - \frac{i}{2}\Gamma_{12}^*}{M_{12} - \frac{i}{2}\Gamma_{12}}} = \frac{M_{12}^* - \frac{i}{2}\Gamma_{12}^*}{\Delta M - \frac{i}{2}\Delta\Gamma}. \tag{11.34}$$

Using the fact that $\Delta M \simeq 2\,\mathrm{Re}M_{12}$ and $\Delta\Gamma \simeq -2\,\mathrm{Re}\Gamma_{12}$, one finds to $\mathcal{O}(\varepsilon)$ that ε can be defined as

$$\varepsilon \simeq \frac{i\;\mathrm{Im}M_{12} + \frac{1}{2}\mathrm{Im}\Gamma_{12}}{\Delta M - \frac{i}{2}\Delta\Gamma}. \tag{11.35}$$

If $\Gamma_{12} \ll \Delta M$, then ε takes the form

$$|\varepsilon| \simeq \frac{\mathrm{Im} M_{12}}{\Delta M}. \tag{11.36}$$

Thus, if M_{12} and Γ_{12} are real, ε will vanish and the states $|K_{S(L)}\rangle$ would correspond to the CP-even(odd) $|K_{1(2)}\rangle$ states. If this is not true and CP is violated, both states are no longer orthogonal

$$\langle K_L|K_S\rangle \approx 2\mathrm{Re}(\bar{\varepsilon}). \tag{11.37}$$

The experimental value of the mass difference, $\Delta M_K = M_{K_L} - M_{K_S}$, is given by [120]

$$\Delta M_K = (3.490 \pm 0.006) \times 10^{-15} \text{ GeV} \tag{11.38}$$

while the experimental value of ε is given by [120]

$$|\varepsilon| = (2.28 \pm 0.02) \times 10^{-3}. \tag{11.39}$$

The measure of K_L decay into two pions through its small CP component was the first observation of CP violation.

11.4.1 The Parameters ε and ΔM_K in the SM

As emphasised, ΔM_K and ε can be calculated from

$$\Delta M_K = 2|\langle K^0|H_{\text{eff}}^{\Delta S=2}|\bar{K}^0\rangle|\,, \tag{11.40}$$

$$|\varepsilon| = \frac{1}{\sqrt{2}\Delta M_K}\mathrm{Im}\langle K^0|H_{\text{eff}}^{\Delta S=2}|\bar{K}^0\rangle\,, \tag{11.41}$$

where $H_{\text{eff}}^{\Delta S=2}$ is the effective Hamiltonian for the $\Delta S = 2$ transition. It can be expressed via the OPE as

$$H_{\text{eff}}^{\Delta S=2} = \sum_i C_i(\mu) Q_i, \tag{11.42}$$

where the $C_i(\mu)$'s are the Wilson coefficients and Q_i's are the relevant local operators. In the SM, the $K^0 - \bar{K}^0$ transition is generated through the $W^{\pm}$ box diagram with up-quark exchanges as shown in Fig. 11.2. Therefore, the effective Hamiltonian is given in terms of a single operator, Q_1, which is defined as

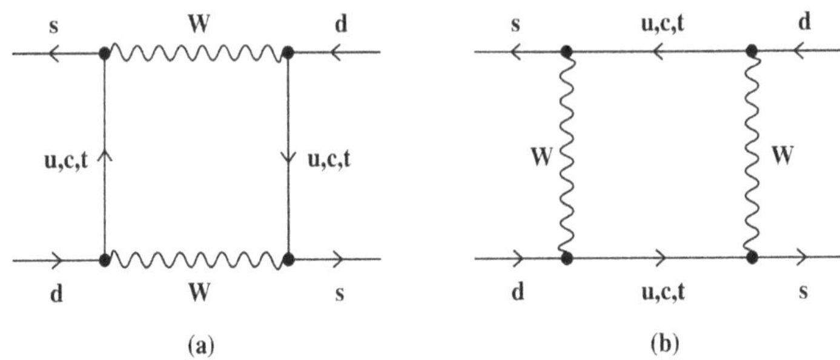

Figure 11.2 SM contribution to $K^0 - \bar{K}^0$ mixing.

$$Q_1 = \bar{d}_L^\alpha \gamma_\mu s_L^\alpha \; \bar{d}_L^\beta \gamma_\mu s_L^\beta \, . \tag{11.43}$$

The corresponding Wilson coefficient is given by [125]

$$C_1^{\rm SM}(M_W) = \frac{G_F^2 M_W^2}{4\pi^2} \left[\eta_1 \lambda_c^2 S_0(x_c) + \eta_2 \lambda_t^2 S_0(x_t) + 2\eta_3 \lambda_c \lambda_t S_0(x_c, x_t)\right] , \tag{11.44}$$

where $\lambda_i = V_{is}^* V_{id}$, η_i are the QCD correction factors with the NLO values $\eta_1 = 1.38 \pm 0.20$, $\eta_2 = 0.57 \pm 0.01$ and $\eta_3 = 0.47 \pm 0.04$, with $S_0(x_i, x_j)$ the loop functions given by [126]

$$S_0(x_c) = x_c \, , \qquad S_0(x_t) = 2.46 \left(\frac{m_t}{173 \rm GeV}\right)^{1.52} , \tag{11.45}$$

$$S_0(x_c, x_t) = x_c \left(\log \frac{x_t}{x_c} - \frac{3x_t}{4(1-x_t)} - \frac{3x_t^2 \log x_t}{4(1-x_t)^2}\right) , \tag{11.46}$$

with $x_i = m_i^2/M_W^2$. The matrix element of the operator Q_1 is parameterised by a term $\hat{B}_1$ as

$$\langle \bar{K}^0 | Q_1 | K^0 \rangle = \frac{1}{3} f_K^2 m_K \hat{B}_1, \tag{11.47}$$

where f_K is the decay width, $f_K \simeq 160$ MeV. Therefore, one finds that $\mathcal{M}_{12}^{\rm SM}$ is given by

$$\mathcal{M}_{12}^{\rm SM} = \frac{G_F^2 M_W^2}{12\pi^2} f_K^2 m_K \hat{B}_K \mathcal{F}^*, \tag{11.48}$$

where

$$\mathcal{F}^* = \eta_1 \lambda_c^2 S_0(x_c) + \eta_2 \lambda_t^2 S_0(x_t) + 2\eta_3 \lambda_c \lambda_t S_0(x_c, x_t). \tag{11.49}$$

Taking $m_c = 1.4$ GeV, $m_t = 173$ GeV as the running quark masses at M_Z, one finds that the SM predictions for ε and ΔM_K are given by:

$$\varepsilon^{\rm SM} \simeq 1.7 \times 10^{-3}, \tag{11.50}$$

$$\Delta M_K^{\rm SM} \simeq 2.26 \times 10^{-15} \text{ GeV}. \tag{11.51}$$

These predictions lie in the ballpark of the measured values. However, a precise prediction cannot be made due to the hadronic and CKM uncertainties. In fact, the main uncertainty in this calculation arises from the matrix elements of Q_i, whereas the Wilson coefficients can be reliably calculated at high energies and evolved down to low energies via the RGE running.

11.4.2 The Ratio ε'/ε in the SM

CP violation can also be induced in direct K decays. In fact, the difference between the decay rate $\Gamma(K^0 \to fX)$ and the CP conjugate $\Gamma(\bar{K}^0 \to \bar{f}X)$ would be a clear indication for CP violation. The parameter ε'/ε that measures direct CP violation in kaons ($\Delta S = 1$) decay is defined as [127]

$$\varepsilon'/\varepsilon = \frac{A(K_L \to (\pi\pi)_{I=2})}{A(K_L \to (\pi\pi)_{I=0})} - \frac{A(K_S \to (\pi\pi)_{I=2})}{A(K_S \to (\pi\pi)_{I=0})}. \tag{11.52}$$

In terms of the amplitudes of two isospin channels in $K^0 \to \pi\pi$, ε' can be approximated as [128]

$$\varepsilon'/\varepsilon \simeq \frac{i}{\sqrt{2}|\varepsilon|} e^{i(\delta_2 - \delta_0)} \text{Im}\frac{A_2}{A_0}, \tag{11.53}$$

where $A_{0,2}$ are the amplitudes for the $\Delta I = 1/2, 3/2$ transitions, defined as

$$\langle(\pi\pi)_I|\mathcal{H}|K^0\rangle = A_I e^{i\delta_I}, \qquad \langle(\pi\pi)_I|\mathcal{H}|\bar{K}^0\rangle = \bar{A}_I e^{i\delta_I}. \tag{11.54}$$

The phases δ_I are the final state interaction (strong interaction) phase, where $\delta_2 - \delta_0 \sim -41°$. The average experimental results of $\text{Re}(\varepsilon'/\varepsilon)$ is given by [129–132]

$$\text{Re}(\varepsilon'/\varepsilon)_{\rm exp} = (1.66 \pm 0.16) \times 10^{-3}, \tag{11.55}$$

which provides firm evidence for the existence of direct CP violation. The effective Hamiltonian for the $\Delta S = 1$ transition is given by [133, 134]

$$H_{\text{eff}}^{\Delta S=1} = \sum_{i=1}^{10} C_i(\mu) Q_i, \tag{11.56}$$

where the C_i's are the Wilson coefficients and $Q_{1,2}$ refer to the current-current operators, Q_{3-6} to QCD penguin operators and Q_{7-10} to EW penguin operators and they are given as follows [134]:

$$Q_1 = (\bar{s}_\alpha u_\beta)_{V-A}(\bar{u}_\beta d_\alpha)_{V-A}, \qquad Q_2 = (\bar{s}u)_{V-A}(\bar{u}d)_{V-A}, \tag{11.57}$$

$$Q_3 = (\bar{s}d)_{V-A}\sum_q (\bar{q}q)_{V-A}, \qquad Q_4 = (\bar{s}_\alpha d_\beta)_{V-A}\sum_q (\bar{q}_\beta q_\alpha)_{V-A}, \tag{11.58}$$

$$Q_5 = (\bar{s}d)_{V-A}\sum_q (\bar{q}q)_{V+A}, \qquad Q_6 = (\bar{s}_\alpha d_\beta)_{V-A}\sum_q (\bar{q}_\beta q_\alpha)_{V+A}, \tag{11.59}$$

$$Q_7 = \frac{3}{2}(\bar{s}d)_{V-A}\sum_q e_q(\bar{q}q)_{V+A}, \qquad Q_8 = \frac{3}{2}(\bar{s}_\alpha d_\beta)_{V-A}\sum_q e_q(\bar{q}_\beta q_\alpha)_{V+A}, \tag{11.60}$$

$$Q_9 = \frac{3}{2}(\bar{s}d)_{V-A}\sum_q e_q(\bar{q}q)_{V-A}, \qquad Q_{10} = \frac{3}{2}(\bar{s}_\alpha d_\beta)_{V-A}\sum_q e_q(\bar{q}_\beta q_\alpha)_{V-A}, \tag{11.61}$$

where α, β are colour indices, e_q are quark charges and $(\bar{f}f)_{V\pm A} \equiv \bar{f}\gamma_\mu(1 \pm \gamma_5)f$. In addition, the operators $\tilde{Q}_i$ are obtained from Q_i by the exchange $L \leftrightarrow R$.

The SM contribution to ε'/ε is dominated by the operators Q_6 and Q_8. They originate from the QCD and EW penguin diagrams and their matrix elements are enhanced by $(m_K/m_s)^2$ [135]:

$$\langle(\pi\pi)_{I=0}|Q_6|K^0\rangle = -4\sqrt{\frac{3}{2}}\left[\frac{m_K}{m_s(\mu)+m_d(\mu)}\right]^2 m_K^2(f_K - f_\pi)\, B_6^{(1/2)}, \tag{11.62}$$

$$\langle(\pi\pi)_{I=2}|Q_6|K^0\rangle = 0, \tag{11.63}$$

$$\langle(\pi\pi)_{I=0}|Q_8|K^0\rangle \simeq 2\sqrt{\frac{3}{2}}\left[\frac{m_K}{m_s(\mu)+m_d(\mu)}\right]^2 m_K^2 f_K\, B_8^{(1/2)}, \tag{11.64}$$

$$\langle(\pi\pi)_{I=2}|Q_8|K^0\rangle \simeq \sqrt{3}\left[\frac{m_K}{m_s(\mu)+m_d(\mu)}\right]^2 m_K^2 f_\pi\, B_8^{(3/2)}, \tag{11.65}$$

where $B_{6,8}^{(1/2)}$ and $B_8^{(3/2)}$ are the so-called 'bag parameters' wherein the matrix elements for the tilde operators come with an opposite sign. In addition, the contributions of the operators Q_6 and Q_8 are enhanced by QCD corrections. Although the Wilson coefficient of Q_8 is suppressed by α/α_s compared to that of Q_6, its contribution to ε' is enhanced by $1/\omega \sim 22$ and is significant. The SM contribution to ε'/ε can be expressed as [135, 136]

$$\varepsilon'/\varepsilon = \text{Im}\lambda_t\; F_{\varepsilon'} \tag{11.66}$$

with

$$F_{\varepsilon'} = P_0 + P_X X_0(x_t) + P_Y Y_0(x_t) + P_Z Z_0(x_t) + P_E E_0(x_t). \tag{11.67}$$

The gauge independent loop functions in the SM are, to a very good approximation, given by

$$X_0(x_t) = 1.57\Big(\frac{m_t}{170\text{GeV}}\Big)^{1.15}, \quad Y_0(x_t) = 1.02\Big(\frac{m_t}{170\text{GeV}}\Big)^{1.56}, \tag{11.68}$$

$$Z_0(x_t) = 0.71\Big(\frac{m_t}{170\text{GeV}}\Big)^{1.86}, \quad E_0(x_t) = 0.26\Big(\frac{m_t}{170\text{GeV}}\Big)^{-1.02}. \tag{11.69}$$

The coefficients P_i are functions of the non-perturbative parameters $B_6^{(1/2)}$ and $B_8^{(3/2)}$:

$$P_i = r_i^{(0)} + \left(\frac{137\text{ MeV}}{m_s(m_c)+m_d(m_c)}\right)^2 \left\{B_6^{(1/2)} r_i^{(6)} + B_8^{(3/2)} r_i^{(8)}\right\}. \tag{11.70}$$

We will use the central values $B_6^{(1/2)} \sim 1.0$ and $B_8^{(3/2)} \sim 0.8$. The numerical factors $r_i^{(x)}$ are given in [136]. Here we use the following values:

$$r_i^0 = -3.122,\ 0.556,\ 0.404,\ 0.412,\ 0.204, \tag{11.71}$$

$$r_i^6 = 10.905,\ 0.019,\ 0.080,\ -0.015,\ -1.276, \tag{11.72}$$

and

$$r_i^8 = 1.423,\ 0,\ 0,\ -9.363,\ 0.409, \tag{11.73}$$

for $i = 0, X, Y, Z$ and E, respectively. The uncertainty in the SM prediction is increased by the fact that there is a large cancellation between the QCD and EW penguin contributions. The estimate of the SM prediction to $\mathrm{Re}(\varepsilon'/\varepsilon)$ leads to [136]

$$\mathrm{Re}(\varepsilon'/\varepsilon)_{\mathrm{SM}} \approx 7.5 \times 10^{-4}. \tag{11.74}$$

This result is particularly sensitive to the bag parameters. Given the theoretical and experimental uncertainties, the SM prediction is consistent with the measured value [129, 137, 138].

11.5 CP VIOLATION IN THE B SYSTEM

There are two neutral $B^0 - \bar{B}^0$ meson systems: $B_q^0 - \bar{B}_q^0$, with $q = d, s$. In these systems, the flavour eigenstates are given by $B_q = (\bar{b}q)$ and $\bar{B}_q = (b\bar{q})$. Like the $K^0 - \bar{K}^0$ system, the $B_q^0 - \bar{B}_q^0$ oscillations are described by the Schrödinger equation:

$$i\frac{d}{dt}\begin{bmatrix} |B_q(t)\rangle \\ |\bar{B}_q(t)\rangle \end{bmatrix} = \left(M^q - \frac{i}{2}\Gamma^q\right)\begin{bmatrix} |B_q(t)\rangle \\ |\bar{B}_q(t)\rangle \end{bmatrix}. \tag{11.75}$$

It is customary to denote the corresponding mass eigenstates by $B_H^q = pB_q + q\bar{B}_q$ and $B_L^q = pB_q - q\bar{B}_q$, where the indices H and L refer to heavy and light mass eigenstates, respectively, and

$$\frac{q}{p} = \left[\frac{\mathcal{M}_{12}^* - \frac{i}{2}\Gamma_{12}^*}{\mathcal{M}_{12} - \frac{i}{2}\Gamma_{12}}\right]^{1/2}. \tag{11.76}$$

Thus, the mass and width differences between B_L^q and B_H^q are given by [139]

$$\Delta M_{B_q} = M_{B_H}^q - M_{B_L}^q = 2|\mathcal{M}_{12}^q|, \tag{11.77}$$

$$\Delta\Gamma_q = \Gamma_L^q - \Gamma_H^q = 2|\Gamma_{12}^q|\cos\phi_q, \tag{11.78}$$

where $\phi_q = \arg\{-\mathcal{M}_{12}^q/\Gamma_{12}^q\}$. Therefore, ΔM_{B_q} can be calculated via

$$\Delta M_{B_q} = 2|\langle B_q^0|H_{\mathrm{eff}}^{\Delta B=2}|\bar{B}_q^0\rangle|, \tag{11.79}$$

where $H_{\text{eff}}^{\Delta B=2}$ is the effective Hamiltonian responsible for the $\Delta B = 2$ transitions.

11.5.1 The Term ΔM_B in the SM

The SM expression for $H_{\text{eff}}^{\Delta B=2}$ is

$$H_{\text{eff}}^{\Delta B=2} = C_1^{\text{SM}} Q_1 + h.c., \tag{11.80}$$

where the four-quark operator Q_1 is given by

$$Q_1 = \bar{q}_L \gamma_\mu b_L \; \bar{q}_L \gamma^\mu b_L \tag{11.81}$$

and the Wilson coefficient C_1^{SM} is defined as

$$C_1^{\text{SM}} = \frac{G_F^2}{4\pi^2} M_W^2 (V_{tq} V_{tb}^*)^2 S_0(x_t). \tag{11.82}$$

The loop function $S_0(x_t)$ of the $\Delta B_q = 2$ box diagram with $W^\pm$ exchange is given in Eq. (11.46). The SM contribution is known at NLO accuracy in QCD. It is given by [134, 139]

$$\mathcal{M}_{12}^{\text{SM}}(B_q) = \frac{G_F^2}{12\pi^2} \eta_B \hat{B}_{B_q} f_{B_q}^2 M_{B_q} M_W^2 (V_{tq} V_{tb}^*)^2 S_0(x_t), \tag{11.83}$$

where f_{B_q} is the B_q meson decay constant, $\hat{B}_{B_q}$ is the RGE invariant B parameter, which is given by 0.87(4), and $\eta_B = 0.8393 \pm 0.0034$. Using these numerical values, one can compute the SM results for ΔM_{B_d}. We find that $(\Delta M_{B_d})^{\text{SM}}$ is given by

$$(\Delta M_{B_d})^{\text{SM}} \approx 0.5 \text{ ps}^{-1}. \tag{11.84}$$

Note that the present experimental value of ΔM_{B_d} is given by

$$(\Delta M_{B_d})^{\text{Exp}} = 0.484 \pm 0.010 \text{ ps}^{-1}. \tag{11.85}$$

Therefore, the small difference between the SM result and the experimental value of ΔM_{B_d} implies stringent constraints on new physics contributions.

Now we turn to $B_s^0 - \bar{B}_s^0$ mixing. It turns out that the SM contribution to ΔM_{B_s} can be estimated more accurately from the ratio $\Delta M_{B_s}^{\text{SM}} / \Delta M_{B_d}^{\text{SM}}$, in which all short-distance effects cancel [139]:

$$\frac{\Delta M_{B_s}^{\rm SM}}{\Delta M_{B_d}^{\rm SM}} = \frac{M_{B_s}}{M_{B_d}} \frac{B_{B_s} f_{B_s}^2}{B_{B_d} f_{B_d}^2} \frac{|V_{ts}|^2}{|V_{td}|^2}. \tag{11.86}$$

The remaining ratio of hadronic parameters has been calculated on the lattice yielding

$$\frac{B_{B_s}(m_b) f_{B_s}^2}{B_{B_d}(m_b) f_{B_d}^2} = (1.15 \pm 0.06^{+0.07}_{-0.00})^2, \tag{11.87}$$

where the asymmetric error is due to the effect of chiral logarithms in the quenched approximation. From the fact that $\Delta M_{B_d}^{\rm SM} \simeq \Delta M_{B_d}^{\rm Exp}$ and $|V_{ts}|^2/|V_{td}|^2$ can be determined from a process which is not constrained by new physics, one finds $\Delta M_{B_s}^{SM} \simeq 15$ ps^{-1} for a quark mixing angle $\gamma \simeq 67°$. The most recent results of ΔM_{B_s} reported by CDF and D0, are given by [140,141]:

$$\Delta M_{B_s} = 17.77 \pm 0.10({\rm stat.}) \pm 0.07({\rm syst.})\ ({\rm CDF}), \tag{11.88}$$

$$\Delta M_{B_s} = 18.53 \pm 0.93({\rm stat.}) \pm 0.30({\rm syst.})\ ({\rm D0}). \tag{11.89}$$

11.5.2 CP Asymmetry in Neutral B-Mesons

When both B^0 and $\bar{B}^0$ decay into CP eigenstate f, the time dependent CP violation asymmetry $A_{\rm CP}$ is defined as

$$A_{\rm CP} = \frac{\Gamma[\bar{B}^0(t) \to f] - \Gamma[B^0(t) \to f]}{\Gamma[\bar{B}^0(t) \to f] + \Gamma[B^0(t) \to f]}, \tag{11.90}$$

where the time evolutions of mesons of B^0 and $\bar{B}^0$ at $t = 0$ are given by

$$|B^0(t)\rangle = g_+(t)|B^0\rangle + \frac{q}{p}\, g_-(t)|\bar{B}^0\rangle, \tag{11.91}$$

$$|\bar{B}^0(t)\rangle = \frac{p}{q}\, g_-(t)|B^0\rangle + g_+(t)|\bar{B}^0\rangle. \tag{11.92}$$

Using the effective Hamiltonian approximation, one can show that

$$g_\pm = \frac{1}{2}\left(e^{-iM_H t - \frac{1}{2}\Gamma_H t} \pm e^{-iM_L t - \frac{1}{2}\Gamma_L t}\right). \tag{11.93}$$

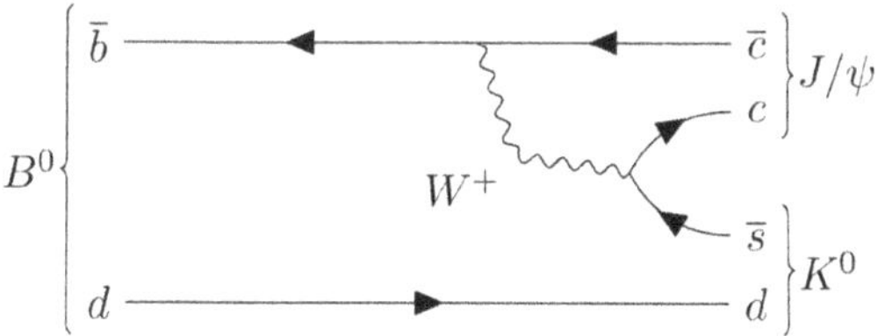

Figure 11.3 The decay topology of $B^0 \to J/\psi K_S^0$.

For $M = (M_H + M_L)/2$ and $\Gamma = (\Gamma_H + \Gamma_L)/2$, $g_\pm(t)$ can be written as

$$g_+(t) = e^{-it(M-\frac{i}{2}\Gamma)} \left[\cosh \frac{\Delta\Gamma}{4} t \cos \frac{\Delta M}{2} t - i \sinh \frac{\Delta\Gamma}{4} t \sin \frac{\Delta M}{2} t\right], \quad (11.94)$$

$$g_-(t) - e^{-it(M-\frac{i}{2}\Gamma)} \left[- \sinh \frac{\Delta\Gamma}{4} t \cos \frac{\Delta M}{2} t + i \cosh \frac{\Delta\Gamma}{4} t \sin \frac{\Delta M}{2} t\right]. \quad (11.95)$$

Using the definition of $\lambda_f = q/p\ \bar{A}_f/A_f$, one finds

$$A_{\rm CP} = \frac{(1 - |\lambda_f|^2)\cos(\Delta M t) - 2{\rm Im}\lambda_f \sin(\Delta M t)}{1 + |\lambda_f|^2}, \quad (11.96)$$

$$= C_f \cos(\Delta M t) - S_f \sin(\Delta M t), \quad (11.97)$$

where we have neglected $\Delta\Gamma$. The last line $S_f \equiv A_{\rm CP}^{\rm mix}$ and $C_f \equiv A_{\rm CP}^{\rm dir}$ are defined as

$$S_f = \frac{2{\rm Im}\lambda_f}{1 + |\lambda_f|^2}, \quad (11.98)$$

$$C_f = \frac{1 - |\lambda_f|^2}{1 + |\lambda_f|^2}. \quad (11.99)$$

Let us consider an interesting example of CP asymmetry in B mesons, namely the decay $B_d \to J/\psi K_S$. The SM contribution to this process is displayed in Fig. 11.3, where the ratio $\bar{A}_{J\psi K_S}/A_{J/\psi K_S}$ is given by

$$\frac{\bar{A}_{J\psi K_S}}{A_{J/\psi K_S}} = \frac{\langle J\psi K_S | H^{\Delta=1} | \bar{B}^0 \rangle}{\langle J\psi K_S | H^{\Delta=1} | B^0 \rangle} = \frac{V_{cs}^* V_c b}{V_{cs} V_{cb}^*} \simeq 1. \quad (11.100)$$

Also q/p will be given by

$$\frac{q}{p} = \sqrt{\frac{M_{12}^*}{M_{12}}} = \frac{V_{tb}^* V_{td}}{V_{tb} V_{td}^*} = e^{-2i\beta}, \tag{11.101}$$

where small Γ_{12} is neglected. Therefore,

$$\lambda_{J/\psi K_S} = \frac{q}{p} \frac{\bar{A}_{J/\psi K_S}}{A_{J/\psi K_S}} = e^{-2i\beta}. \tag{11.102}$$

Hence

$$S_{J/\psi K_S}^{\rm SM} = \sin 2\beta, \qquad C_{J/\psi K_S}^{\rm SM} = 0. \tag{11.103}$$

From the UTfit result [122] for the angle β, it turns out the SM results for direct and mixing CP asymmetries are very consistent with the experimental measurements, which are given by

$$S_{J/\psi K_S} = 0.73 \pm 0.07(\text{stat.}) \pm 0.04(\text{syst.}), \tag{11.104}$$

$$C_{J/\psi K_S} = 0.03 \pm 0.09(\text{stat.}) \pm 0.01(\text{syst.}), \tag{11.105}$$

CHAPTER 12

Tests of EW Interactions

In this chapter, we will review the phenomenological progress made in establishing the weak interactions of the SM. We will start by describing the theoretical situation when such interactions were detected for the first time. We will then move on to illustrate how their discovery took place. Finally, we will discuss various precision measurements carried out at various colliders between the early nineties and into the first decade of the new century for the purpose of testing the weak interactions experimentally.

12.1 WEAK INTERACTIONS FROM UNITARITY

As already mentioned, weak interactions were discovered in β-decay and correctly described by the effective Lagrangian that we have previously discussed within so-called 'Fermi theory'. If such a description is applied to muon (instead of neutron) decay, *i.e.*, $\mu^- \to e^- \bar{\nu}_e \nu_\mu$, it reads as:

$$\mathcal{L} = \frac{G_F}{\sqrt{2}} [\bar{\nu}_\mu \gamma_\lambda (1 - \gamma_5) \mu][\bar{e} \gamma^\lambda (1 - \gamma_5) \nu_e], \tag{12.1}$$

with $G_F \approx 1.17 \times 10^{-5}$ GeV^{-2} being the usual Fermi coupling. In essence, this Lagrangian describes an effective low-energy theory and cannot be extended to arbitrarily high energies. In fact, if one applies such an effective Lagrangian at high energies, *e.g.*, to $\bar{\nu}_\mu e^- \to \mu^- \nu_e$ scattering, one obtain the relevant scattering ME, this reads as

$$\mathcal{M}[\bar{\nu}_\mu e^- \to \mu^- \nu_e] \sim \frac{G_F s}{2\sqrt{2}\pi}. \tag{12.2}$$

DOI: 10.1201/9780429443015-12

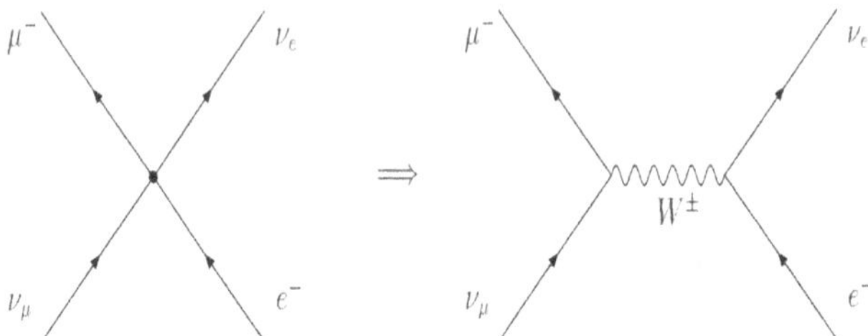

Figure 12.1 Representation of $\bar{\nu}_\mu e^- \to \mu^- \nu_e$ scattering in an effective field theory (left) and in a quantum field theory (right).

Now, in a well-defined theory, such a scattering amplitude must respect the following unitarity bound:

$$|\text{Re}\mathcal{M}| \leq 1/2. \tag{12.3}$$

It therefore follows that the Fermi theory cannot be applied when $s \gtrsim (600 \text{ GeV})^2$. This breakdown of unitarity signals that new states are necessary. Indeed, such a unitarity problem in the Fermi theory is elegantly solved by assuming that weak interactions are mediated by massive charged vector bosons, as depicted in Fig. 12.1. Thus, the introduction of a $W^\pm$ propagator connecting the initial and final states cures this problem, as it dampens the rise of the scattering amplitude as $\sqrt{s} \to \infty$,

$$\mathcal{M}[\bar{\nu}_\mu e^- \to \mu^- \nu_e] \to \frac{G_F s}{2\sqrt{2}\pi} \frac{M_W^2}{M_W^2 - s}, \tag{12.4}$$

if $M_W \approx 100$ GeV, in turn giving a clear indication on the mass scale where NP will appear (in this case, in the form of a new massive charged gauge boson, as previously remarked).

In fact, rather intriguingly, one can deduce most of the structure of weak interactions from imposing such unitarity constraints, as pursued by Llewellyn-Smith [142] as well as Cornwall, Levin and Tiktopoulos [143, 144] (see also [145]). For example, consider the production of W^+W^- pairs in e^+e^- annihilations. The neutrino term (stemming from the left diagram in Fig. 12.2) grows quadratically and violates unitarity. Hence, this hints at the fact that there should be other interactions mediating such a processes, as per the right diagram in Fig. 12.2. Indeed, the bad high-energy behaviour of the neutrino diagram is cured by the exchange of one (or more) new neutral vector boson W^3 in s-channel, which cannot be a photon, as the latter would not induce a divergent behaviour with s. Needless to say, this was a clear hint of the existence of what would have eventually been identified as the Z boson, which, in order to onset an effective cancellation, must have a mass similar to that of the $W^\pm$ boson [146–149].

So, by simply requiring the Fermi theory to be unitary, we have been able to explain the existence of the $W^\pm$ and Z states, which is precisely what the Glashow-Weinberg-Salam model eventually predicted. But let us continue in our quest to map the entire EW sector of the SM by resorting to this requirement of self-consistency of the theory. Consider now $WW \to WW$ scattering, as exemplified in Fig. 12.3, which captures all weak interactions known to us, via the left diagram (which in fact comes in all possible permutations), plus a new one, the right diagram, which is indeed expected on the basis of gauge invariance and power counting. Again, the former grows quadratically with energy as $s \to \infty$,

$$\mathcal{M}[WW \to WW] \sim \frac{G_F s}{4\sqrt{2}\pi}, \tag{12.5}$$

so that, for self-consistency, such a growth must be dampened. If the theory is to satisfy unitarity, a new interaction must take place, like the one sketched in Fig. 12.4, wherein a massive neutral scalar particle is introduced, which couples to a particle with a strength proportional to its mass. Such a new state, indeed a Higgs boson, generates interactions that cancel the above bad high-energy behaviour yielding

$$\mathcal{M}[WW \to WW] \sim -\frac{G_F M_H^2}{4\sqrt{2}\pi}, \tag{12.6}$$

so that the complete amplitude fulfils the unitarity requirement, so long that $M_H \lesssim 1$ TeV [150, 151]. Therefore, long before EWSB had been theoretically formalised through the Higgs mechanism as described in previous chapters, a clear hint for the ensuing Higgs boson had been gathered again from the requirement of unitarity of the underlying theory of weak interactions.

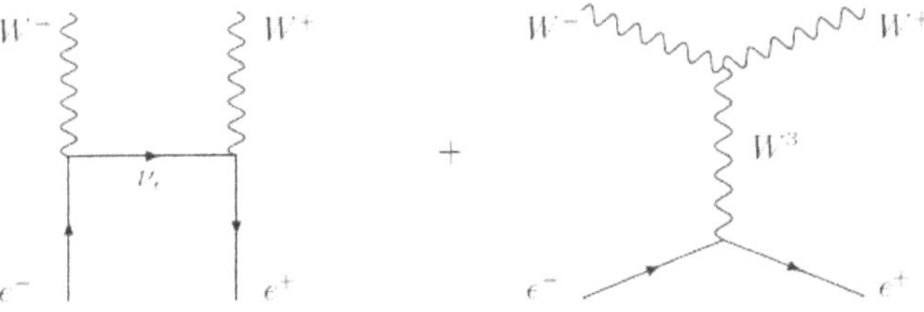

Figure 12.2 The two contributions to $e^+e^- \to W^+W^-$ scattering: ν_e exchange (left) and neutral vector boson exchange (right).

12.2 DISCOVERY OF THE $W^\pm$ AND Z PARTICLES

While predicted by theory since the 1960s, the production of the $W^\pm$ and Z particles (as real objects) was beyond the reach of existing apparata of the time. In fact, the β-theory (and a myriad of other experiments) involved the production of virtual gauge bosons and specifically only of the charged form, as the predicted neutral current interactions (the exchange of Z bosons) had never been observed. In the latter respect, a step-change occurred in 1973, with the observation of neutral current interactions as predicted by the EW theory. The large Gargamelle bubble chamber photographed the tracks of a few electrons suddenly starting to move, seemingly of their own accord [152, 153]. This was interpreted as a neutrino interacting with the electron by the exchange of an unseen (virtual) Z boson. The neutrino is otherwise undetectable, so the only observable effect was the momentum imparted to the electron by the interaction.

The actual production of real $W^\pm$ and Z bosons and their detection required huge investments in the construction of extremely large particle accelerators and detectors, which could only be attained in a few high-energy physics laboratories in the world (and then only starting from the late seventies), so that the discovery of the $W^\pm$ and Z bosons was claimed at CERN only in the eighties, when such apparata had become powerful enough, as we have seen that their masses were expected to be at the 100 GeV or so scale (*i.e.*, 100 times heavier that the nucleons entering β-decay). Indeed, neither the ISR pp collider ($\sqrt{s} = 61$ GeV) nor the SPS proton accelerator for fixed target experiments (wherein $E_p = 400$ GeV was also not sufficient to produce

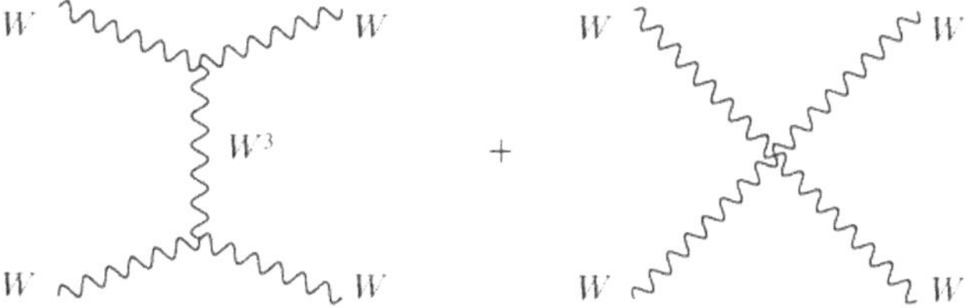

Figure 12.3 The two gauge contributions to $WW \to WW$ scattering: W^3 exchange (left) and quartic interaction (right).

enough energy in the centre-of-mass, $\sqrt{s} = \sqrt{2mE_p}$, given the target mass $m \sim 1$ GeV) at CERN were powerful enough to succeed in this quest. Eventually, the upgrade of the SPS into the S$p\bar{p}$S, strongly driven by Rubbia and van der Meer, enabled to attain $\sqrt{s} = 540$ GeV by December 1981. Herein, unambiguous signals of $W^\pm$ bosons were seen in January 1983 by two experiments called UA1 (led by Rubbia) and UA2 (led by Darriulat), see Fig. 12.5. Finally, UA1 and UA2 found the Z boson a few months later, in May 1983. As both the $W^\pm$ and Z boson are highly unstable, only their decay products, $\ell^\pm\nu_\ell$ and $\ell^+\ell^-$ ($\ell = e, \mu$), respectively, were actually seen.

12.3 PRECISE DETERMINATIONS OF EW OBSERVABLES

As we have seen, well in the spirit of the phenomenological approach described in our Preamble, a twofold approach was developed in testing EW interactions. On the one hand, EW interactions neatly emerged from enforcing the requirement of unitarity at high energies. On the other hand, the SM was formulated in terms of a Yang-Mills theory. As explained, in the initial chapters of the book, the latter is a gauge theory based on special unitary groups $SU(N)$, or more generally any compact reductive Lie algebra, which is able to describe the behaviour of elementary particles using these non-Abelian Lie groups and is thus at the core of the unification of the EM and weak forces into the EW one (based on $U(1)_Y \times SU(2)_L$) as well as QCD (based on $SU(3)_C$). In a sense, then, it forms the basis of our understanding of the SM of particle physics.

The possibility of describing EW and QCD interactions involving matter in terms of Yan-Mills theories was beautifully affirmed by 't Hooft in 1972,

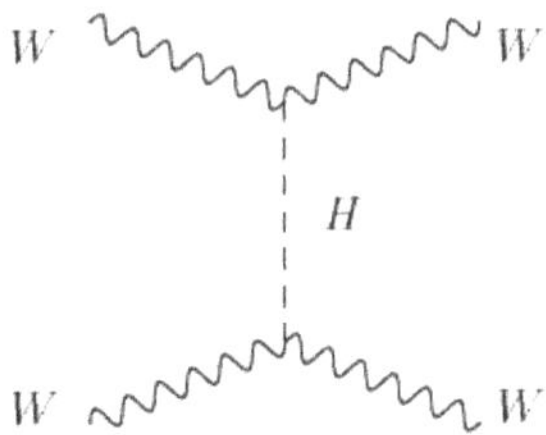

Figure 12.4 The Higgs boson contribution to $WW \to WW$ scattering.

when he worked out their renormalisability, following on from initial work by Veltman. While the theory of renormalisation was well known from QED, for which it was first developed, the crucial difference is that 't Hooft and Veltman proved that renormalisability is obtained in Yang-Mills theories even if the gauge bosons described by these are massive, provided the mass is spontaneously generated, as it happens through the Higgs mechanism within the SM. The essence of this procedure is that SM observables can be calculated to arbitrarily high precision in a systematic expansion after a few basic parameters are fixed experimentally. In fact, quantum corrections in interacting field theories modify particle masses and couplings[1], *i.e.*, such interactions renormalise the fundamental parameters of the SM. Such quantum corrections are described by Feynman diagrams including loops, like in Fig. 12.6.

These corrections have the following typical energy dependence:

$$\sim \int \frac{d^4k}{k^4} \sim \ln \Lambda^2_{\mathrm{cut}}. \tag{12.7}$$

The so-called self-energy corrections (to the mass) and vertex corrections (to the coupling) are thus logarithmically divergent for large loop momenta (hence they are called Ultra-Violet (UV) corrections) and lead to contributions $\sim \ln \Lambda^2_{\mathrm{cut}}$, where Λ_{cut} is the energy scale up to which the SM is valid, precisely the same scale discussed in Chapter 9. Such quantum corrections add to the unobservable 'bare mass' m_0 and 'bare coupling' g_0 to generate the observable physical mass m and coupling g, *i.e.*, the latter are obtained as $m_0 + \delta m = m$ and $g_0 + \delta g = g$. The bare terms and quantum corrections are

[1]The same argument can be made for wavefunctions; however, being less intuitive, we leave this aside.

individually divergent, but they cancel to produce finite results, *i.e.*, renormalisation is sufficient to absorb all divergences and render all masses and couplings (upon which other physics observables depend) finite if $\Lambda_{\text{cut}} \to \infty$. The arbitrariness in fixing the value of the bare quantities is finally removed once masses m and couplings g are fixed experimentally to some measured values for each particle species, so that all ensuing observables are calculable to arbitrarily high precision. Such a renormalisation procedure is process independent, *i.e.*, it does not depend upon through which physics process is performed, and such renormalised quantities can be used ubiquitously to make meaningful theoretical predictions. Indeed, the precision of the latter is quite simply limited by our technical abilities in computing them.

A complete understanding of EW interactions in higher order in perturbation theory is crucial in order to test the SM in high precision experiments, following the discovery of the $W^\pm$ and Z bosons. In fact, the next stage in accelerator development in such a pursuit involved the construction of leptonic colliders, such as the e^+e^- LEP collider at CERN (a circular machine) and the SLC at SLAC (a linear machine). When run at $\sqrt{s} = M_Z$ (the energy of the first phase of LEP, so-called LEP1, and SLC), such colliders can produce copiously the neutral weak boson Z, via the diagram in Fig. 8.2 (where γ^* mediation is supplemented by Z exchange), as the cross section would see a significant Breit-Wigner enhancement as the one seen in Fig. 8.3. Now, consider the generic process $e^+e^- \to f\bar{f}$ (with $f = q, \ell, \nu$, where ℓ now can include τ's and ν refers to a neutrino) in the SM, for which one can write the γ and Z terms of the cross section associated with each fermion f as follows:

$$\sigma_\gamma(s) = \frac{4\pi\alpha^2}{3s} Q_f^2 N_f \qquad (N_q = N_C,\ N_{\ell,\nu} = 1), \tag{12.8}$$

$$\sigma_Z(s) = \frac{4\pi\alpha^2}{3s} \frac{s^2}{(s - M_Z^2)^2 + M_Z^2\Gamma_Z^2} \mathcal{A}_f \mathcal{A}_e N_f, \tag{12.9}$$

with

$$\mathcal{A}_f = v_f^2 + a_f^2 = \frac{(t_{3f} - 2Q_f \sin^2\theta_W)^2 + t_{3f}^2}{4\sin^2\theta_W \cos^2\theta_W}, \tag{12.10}$$

with v_f and a_f the vector and axial coupling, respectively, of the Z to the fermion f.

Clearly, as the photon term was well known, LEP1 and SLC had the possibility of accessing the Z couplings to all fermion species, thereby verifying that

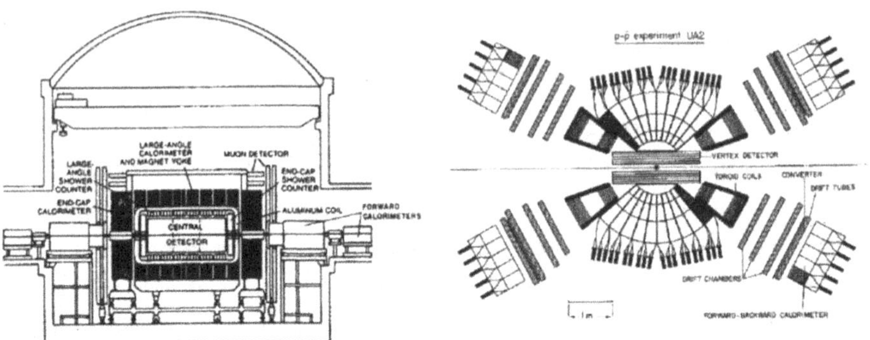

Figure 12.5 Schematic view of the UA1 (left) and UA2 (right) detectors.

they indeed had the properties described by the Glashow-Weinberg-Salam theory. In reality, the procedure is somewhat more complicated by the fact that there is an interference[2] between the γ and Z terms (which only vanishes for $\sqrt{s} = M_Z$), so that the total cross section for each fermion flavour reads as

$$\begin{aligned}\sigma_{\rm SM}(s) &= \frac{4\pi\alpha^2(s)}{3s}\frac{s^2}{(s-M_Z^2)^2+(s^2/M_Z^2)\Gamma_Z^2}\Big[1+\underbrace{\Delta_Z}_{\displaystyle\downarrow}\Big]\mathcal{A}_f\mathcal{A}_e N_f \\ &+ \frac{4\pi\alpha^2(s)}{3s}Q_f^2 N_f, \qquad\qquad (\gamma - Z \text{ interference})\end{aligned} \tag{12.11}$$

where we also have made clear that the EM coupling constant runs, $\alpha \to \alpha(s)$, the so-called "improved Born approximation", which captures the leading logarithmic radiative corrections due to (both real and virtual) photon exchange. In fact, not only higher order EM effects ought to be included in the theoretical predictions, but also purely weak ones, such as those in Fig. 12.7, which, as explained, lead to ultraviolet divergences which have to be absorbed into renormalised masses (left) and couplings (right). In addition, QCD corrections have to be included in the case $f = q$, as explained in Chapter 8.

In fact, it may be very instructive to take another closer look at the hadronic cross section in the vicinity of the Z mass peak, as shown in Fig. 12.8. Even for such an observable, which, as explained previously, is one of the least precise amongst those which can be investigated experimentally, owing to the QCD corrections being much larger than the EW ones (at least at

[2]Which expression we do not write down explicitly here at it is rather lengthy.

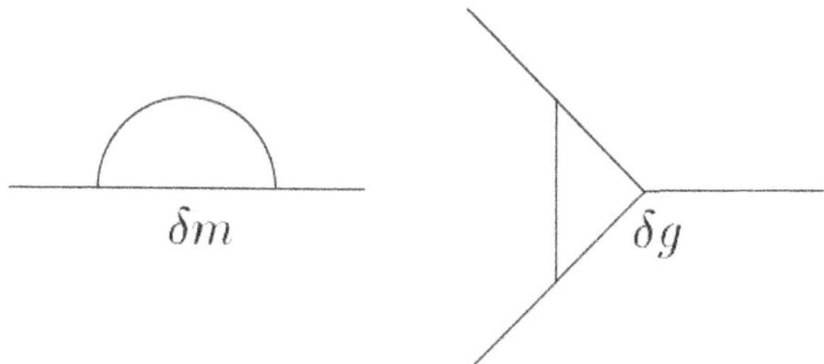

Figure 12.6 Feynman diagram description of quantum corrections to the mass m (left) and coupling g (right).

LEP energy scales), it is remarkable to see how well theory and experiment coincide. It should in fact be appreciated that, herein, the measured cross section (solid curve) is somewhat shifted with respect to the one predicted by the SM (dashed curve), because the ISR of photons emitted by the incoming electron-positron pair reduces the effective energy available to the annihilation process. However, this effect can be modelled extremely precisely in the QED theory so that, ultimately, such a measurement can for example be used in extracting both M_Z and Γ_Z via a fit to actual data. The still unrivalled LEP1 measurements of these two quantities are: $M_Z = 91.1876 \pm 0.0021$ GeV and $\Gamma_Z = 2.4952 \pm 0.0023$ GeV.

Equipped with the measurements of α, G_F, M_Z and Γ_Z, one can perform a first non-trivial test of the EW interactions at the Z peak, by extracting $\mathcal{A}_f$ and $\mathcal{A}_e$ from the following expression:

$$\sigma_{\rm SM}(M_Z) = \frac{4\pi\alpha^2}{3\Gamma_Z^2}\mathcal{A}_f\mathcal{A}_e. \tag{12.12}$$

Furthermore, it is possible to compute the value of Γ_Z as predicted by the SM, through the contribution of each final state fermion f, which reads at tree-level as

$$\Gamma_f = \frac{1}{3}\alpha\, M_Z\, \mathcal{A}_f. \tag{12.13}$$

Hence, the total Z width is obtained from a summation of all possible final states (*i.e.*, those with quarks, leptons and neutrinos with mass less than

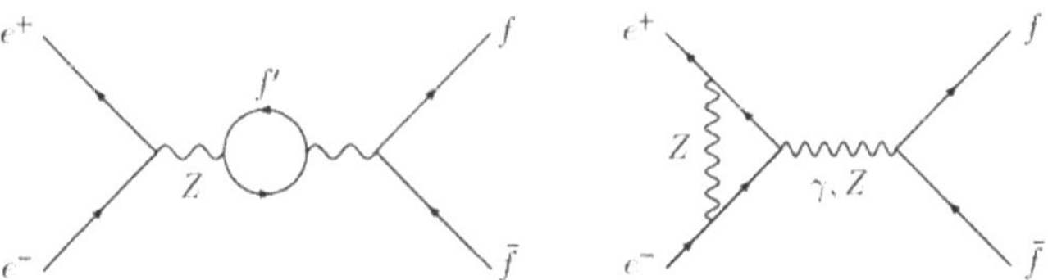

Figure 12.7 Feynman diagram contribution to the Z mass (left) and coupling (right) corrections.

$M_Z/2$),

$$\Gamma_Z = \sum_f \Gamma_f = \sum_{\ell=1}^{N_\ell} \Gamma_\ell + \sum_{\nu=1}^{N_\nu} \Gamma_\nu + \sum_{q=1}^{N_q} \Gamma_q, \tag{12.14}$$

which was then compared to the measured value (combined between LEP1 and SLC) of $\Gamma_Z = 2.4952 \pm 0.0023$ GeV, thus enabling a further non-trivial test of the SM. The confrontation of theory and experiment in this observable also provides an independent measurement of the number of neutrino species: $N_\nu = 2.993 \pm 0.011$ (given that $N_\ell = 3$ and $N_q = 5$ were known). Conversely, in the presence of the finite experimental error on Γ_Z, one could place a limit on the width to additional invisible particles: $\Gamma_{\text{invisible}} = 499.0 \pm 1.5$ MeV.

Notice, though, that cross section measurements, through the Z line-shape and width, are only sensitive to combinations of $\mathcal{A}_f = v_f^2 + a_f^2$. However, the differential cross section also contains a dependence on $\cos\theta$, the polar angle of the outgoing fermion, which coefficient reads as follows:

$$\mathcal{B}_f = 2v_f a_f. \tag{12.15}$$

Therefore, one can construct the so-called Forward-Backward Asymmetry (A_{FB}) of the cross section:

$$\begin{aligned} A_{\text{FB}} &\equiv \frac{\sigma_{\text{SM}}(\theta < 90^\circ) - \sigma_{\text{SM}}(\theta > 90^\circ)}{\sigma_{\text{SM}}(\theta < 90^\circ) + \sigma_{\text{SM}}(\theta > 90^\circ)} \\ &= \frac{3}{4}\frac{\mathcal{B}_e \mathcal{B}_f}{\mathcal{A}_e \mathcal{A}_f}, \end{aligned} \tag{12.16}$$

where the polar angle can unambiguously be identified for each event since the directions of, *e.g.*, the incoming and outgoing fermion (or antifermion) are

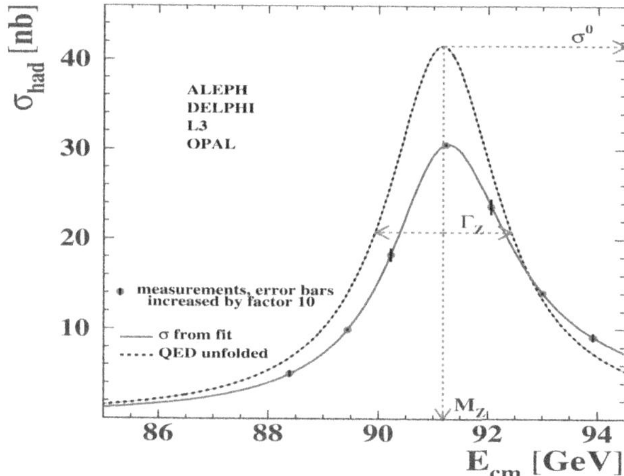

Figure 12.8 The $e^+e^- \to$ hadrons cross section in the vicinity of $\sqrt{s} = M_Z$ as predicted by the SM and measured by the LEP Collaborations (ALEPH, DELPHI, L3 and OPAL). (From Ref. [154].)

always known, except for neutrinos[3]. This offered the possibility of new, and complementary, tests of the SM.

Another asymmetry can be exploited in e^+e^- annihilations if either beam can be polarised and SLC indeed had this feature, as it could produce highly polarised electrons ($P_{e^-} \sim 69\%$). This is called the Left-Right Asymmetry ($A_{\rm LR}$), defined as follows:

$$A_{\rm LR} \equiv \frac{\sigma_{\rm SM}(e^+e^-_L) - \sigma_{\rm SM}(e^+e^-_R)}{\sigma_{\rm SM}(e^+e^-_L) + \sigma_{\rm SM}(e^+e^-_R)} = -\frac{\mathcal{B}_e}{\mathcal{A}_e} \tag{12.17}$$

and has three notable advantages: (i) it is independent of the final state, (ii) it is independent of the angular range, (iii) it is much larger than $A_{\rm FB}$. Its key feature is therefore the high sensitivity to the initial state parameters through an almost systematically error-free measurement, so long that the polarisation is well measured. Unsurprisingly then, the SLD Collaboration provided the world best measurement of $\sin^2\theta_W$.

In order to probe the $W^\pm$ properties in e^+e^- annihilations, it was necessary to produce it in pairs. This happened at the second stage of LEP, which ran

[3]Also for quarks, as it is possible to measure the jet EM charge to extract that of the parton.

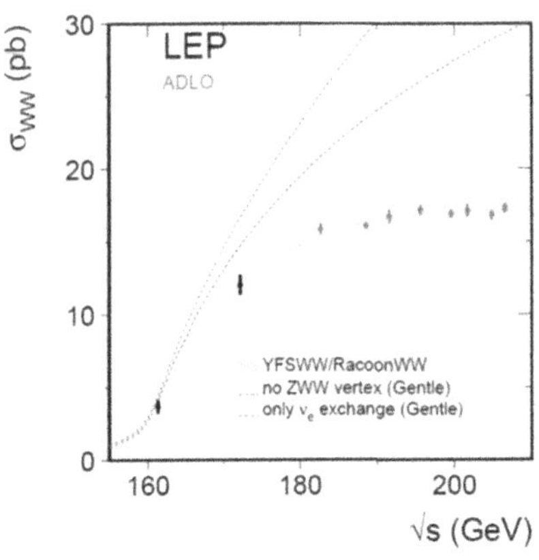

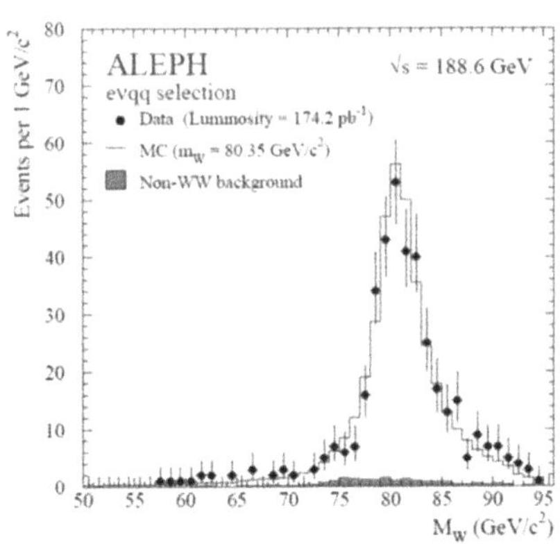

Figure 12.9 (Left) The $e^+e^- \to W^+W^-$ cross section as a function of the collider energy. (Right) The reconstructed $W^\pm$ mass from the di-jet system in semileptonic decays of W^+W^- pairs at fixed $\sqrt{s}$.

at energies from just below $2M_W$ up to 209 GeV. The Feynman diagrams enabling this are those in Fig. 12.2, wherein in the SM one has $W^3 = \gamma, Z$. As previously explained, the presence of the Z exchange cures the divergent behaviour of neutrino exchange, since each of the two contributions diverges like

$$\mathcal{M}[e^+e^- \to W^+W^-] \sim \frac{G_F^2 s}{48\pi} \qquad (s \gg M_W^2) \tag{12.18}$$

while the sum has a tamed asymptotical behaviour,

$$\mathcal{M}[e^+e^- \to W^+W^-] \sim \frac{G_F^2 M_W^4}{s\pi} \log \frac{s}{M_W^2} \qquad (s \gg M_W^2), \tag{12.19}$$

so that, in turn, one is extremely sensitive to the value of the new TGC, ZW^+W^-, which is indeed predicted within the SM[4].

Also the measurement of M_W at LEP2 would have enabled another strong test of the SM as its value is predicted by it once α, G_F and M_Z are measured, specifically, of the symmetry breaking mechanism itself. Such a parameter was measured in two different ways, from a fit to the total cross section of

[4]Recall that the γW^+W^- coupling is fixed by the QED Ward identity [42].

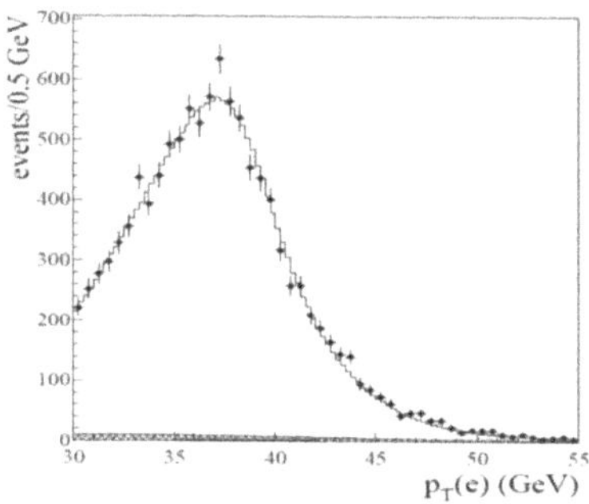

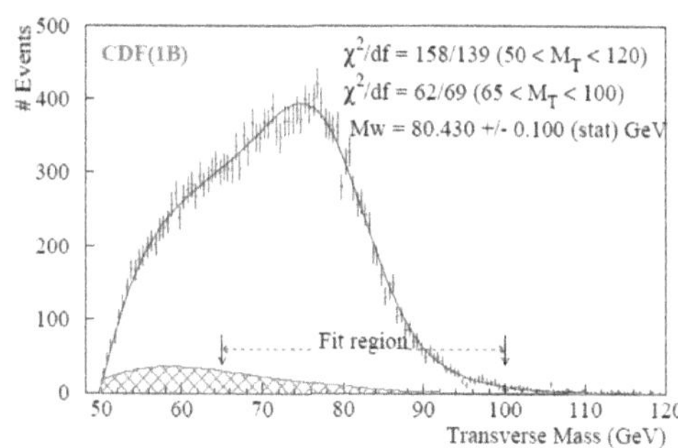

Figure 12.10 Distributions of $W^{\pm} \to e\nu$ events at the TeVatron: the lepton transverse momentum from D0 (left) and the transverse mass from CDF (right).

$e^+e^- \to W^+W^-$ near the threshold at $2M_W$, where one has

$$\sigma_{WW} \sim \frac{G_F^2 M_W^2}{2\pi} \underbrace{\sqrt{1 - \frac{4M_W^2}{s}}}_{\substack{\text{velocity of } W \\ \text{rapidly varying for } \sqrt{s} \sim 2M_W}}, \tag{12.20}$$

which is very clean theoretically (*i.e.*, with a very small systematical error) but experimentally suffers from large statistical uncertainties (owing to the cross section being rather small near threshold), see Fig. 12.9 (left). A second method requires one to measure, at a fixed collider energy, the invariant mass of the $W \to$ jet decays in events initiated by, *e.g.*, $e\nu q\bar{q}$ final states, see Fig. 12.9 (right). By the end of LEP2, the measurement of the $W^{\pm}$ mass read as $M_W = 80.412 \pm 0.042$ GeV.

While the CERN collider was in operation, another one concurrently ran at FNAL, the TeVatron (a pp collider with $\sqrt{s}$ from 1.8 to 1.96 TeV), which was also able to produce $W^{\pm}$ bosons, on the same footing as the $Sp\bar{p}S$ did some decades earlier, albeit with much higher cross sections, *i.e.*, via the process $q\bar{q}' \to W^{\pm} \to \ell\bar{\nu}_\ell/\bar{\ell}\nu_\ell$, with $\ell = e, \mu$ decays providing a small but clean sample. As the neutrinos are lost to the detectors, so that only their transverse momentum p_T^ν can be used (it equates to the missing one), two observables are used in the $W^{\pm}$ mass reconstruction: the lepton transverse momentum $p_T(\ell)$ (see the left frame of Fig. 12.10) and the so-called transverse mass $M_T^2 \equiv 2p_T^e p_T^\nu(1 - \cos\phi)$ (see right frame of Fig. 12.10). The final measurement from the FNAL collider was $M_W = 80.452 \pm 0.059$ GeV, hence very competitive with the CERN one. The final combination obtained after both machines had ended operations read as $M_W = 80.425 \pm 0.034$ GeV. Eventually, sensitivity to Γ_W was also established, giving the combined measurement $\Gamma_W = 2.147 \pm 0.060$ GeV.

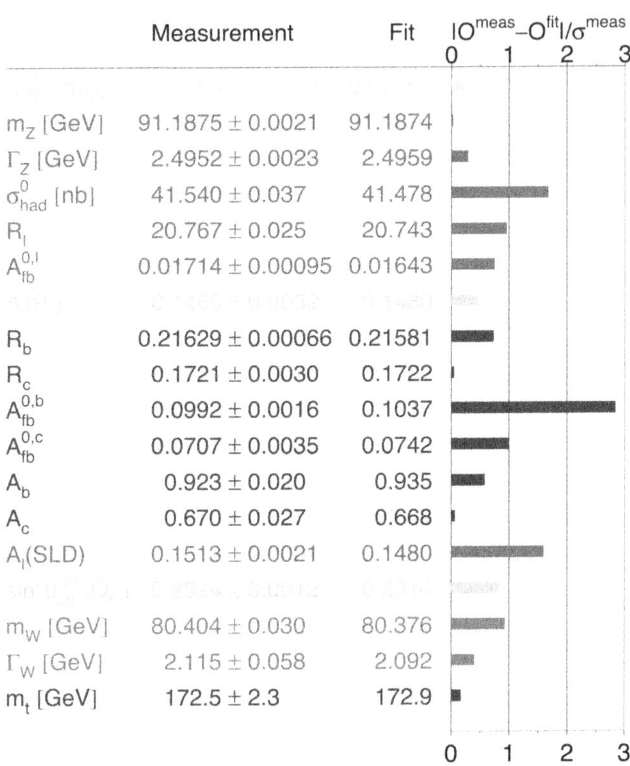

Figure 12.11 The pull (as defined in the text) of a variety of EW precision measurements as obtained by the LEPEWWG in the Winter of 2006.

One of the latest compilations of the EW precision measurements described in this section, as well as many more, as produced by the LEPEWWG, which combined results from not only ALEPH, DELPHI, L3 and OPAL but also NuTeV, CDF, D0 and SLD (see Ref. [154], to which we refer for their definitions) can be found in Fig. 12.11, which displays the so-called "pull", which is defined as the deviation of data from the theoretical prediction in units of the corresponding one-standard deviation experimental uncertainty. The agreement between the SM predictions and all such data is remarkable[5]. However, the attentive reader will have noticed also the presence of the top-quark mass in this table and may wonder how this could be. This is a very interesting story, which deserves a new chapter.

[5]See Ref. [155] for their most up-to-date values, in the presence of more recent measurements from other experiments.

CHAPTER 13

Top and Higgs Discovery

In this chapter, we will review other two remarkable success stories of the SM. They both in part stem from the same physics as in the previous chapter, as the high precision measurements of EW observables paved the way to the discovery of the last two building blocks of such a theoretical construction, namely, the top quark and Higgs boson. In the sense that all such measurements where fully consistent with the parameters (masses and couplings) of these two particles hitherto undetected, when their presence was allowed for in the theoretical predictions. The actual discovery of the top quark and Higgs boson was however achieved quite irrespectively of such measurements, thanks to the efforts of the physics communities working at FNAL and CERN, respectively, which were responsible for first detecting these objects.

13.1 INDIRECT SEARCH FOR TOP AND HIGGS

As we have remarked in the previous chapter, precision observables are affected by quantum corrections within the theory, to which various experiments can be sensitive. Already at the one-loop level, such corrections carry a dependence on all particle entering a theoretical model, even those which mass is much higher than the energy scale at which such experiments are performed, since they can then be produced as virtual objects, rather than real ones. Within the framework of the SM, one therefore realises that loop computations carry a dependence also on states which were not discovered yet at the time these high precision experiments (chiefly, of LEP and SLC) were carried out, indeed, the top quark and Higgs bosons. As their couplings to other SM states were fixed by the SM, such data gave access to the two new high mass scales of the

DOI: 10.1201/9780429443015-13

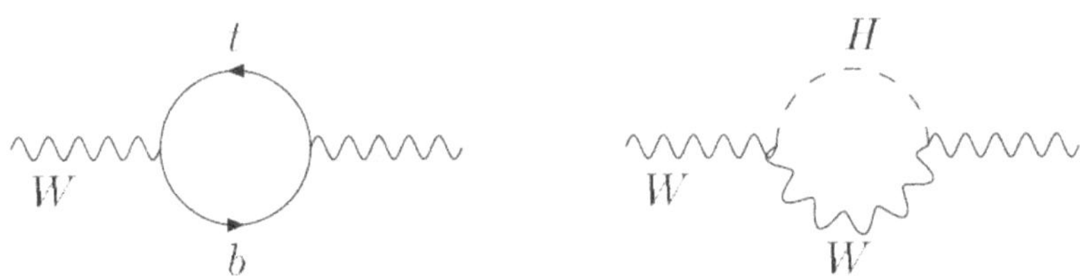

Figure 13.1 The dependence at one-loop level in the SM of the $W^{\pm}$ mass on those of the top quark and Higgs boson.

SM: m_t and M_H, as t and H enter in many loop corrections to several EW observables. One of the most sensitive measurements is that of the so-called ρ parameter, defined as

$$\rho = \frac{M_W^2}{M_Z^2 \cos\theta_W^2}, \tag{13.1}$$

which is equal to 1 at tree-level in the SM. This observable captures the three fundamental parameters of the weak interactions, *i.e.*, M_W, M_Z and $\sin^2\theta_W$ and their relation. Now, consider, *e.g.*, the $W^{\pm}$ mass, which in the SM is predicted as being the pole of its propagator, the latter has one-loop contributions like those exemplified in Fig. 13.1, indeed carrying a dependence on both the top quark and Higgs boson mass. The key feature here is that such a dependence is quadratic on m_t (*i.e.*, $\propto m_t^2$) and logarithmic on M_H (*i.e.*, $\propto \log M_H$). Therefore, the sensitivity is largest in the case of the former than that of the latter. This is well exemplified by the plot in Fig. 13.2. This displays the region allowed by LEP, SLD and TeVatron data on the (M_H, m_t) plane within one standard deviation from the measured means (area inside the blue contour). Owing to the aforementioned different dependence upon m_t and M_H, it is not surprising to see that the while the top quark mass is constrained within a few GeV that of the Higgs boson is limited over more than one order of magnitude.

Regarding this last figure, we are sure that the attentive reader will have noticed the presence of two bands on the plots. The first one corresponds to the direct measurements of m_t (horizontal band) of width ± 1 standard deviation as eventually obtained at the TeVatron from the discovery of the top quark by CDF [23, 156] and D0 [24] while the vertical band shows the 95% CL exclusion limit on M_H up to 114 GeV as derived from direct searches for the Higgs boson at LEP2 by all four Collaborations (ALEPH, DELPHI, L3 and OPAL) [157]. However, before moving on to describe the direct searches for these objects, let us first amaze ourselves with Fig. 13.3, which makes the point that we intimated at the beginning of this chapter, that EW precision observables led to a prediction for m_t. While some controversy still exists to date on how the e^+e^- and pp experiments influenced each other in the quest for the top quark, it is remarkable to compare the values obtained by the time the former had finished analysing their full datasets with those that the latter were still performing. In the PDG compilation of Ref. [158] the two numbers read as follows:

$$m_t = 174.3 \pm 5.1 \text{ GeV (direct observation)},$$

$$m_t = 178.1^{+10.4}_{-8.3} \text{ GeV (SM EW fits)}.$$

This represents a remarkable success of the SM as a renormalisable theory so it is perhaps not surprising to note here that it took till 1999 to the Nobel Foundation to award the Nobel Prize in Physics to 't Hooft and Veltman.

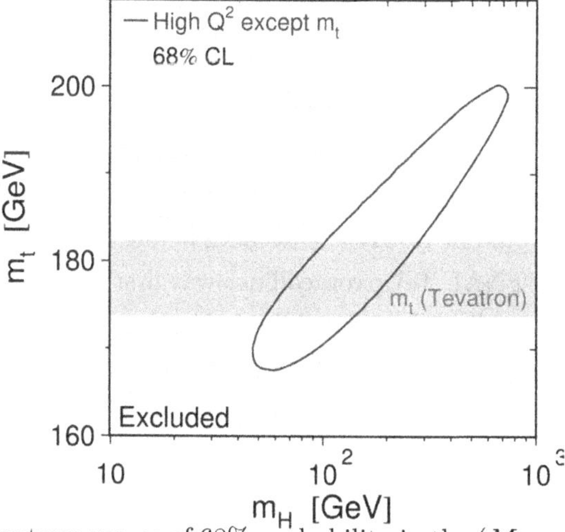

Figure 13.2 Contour curves of 68% probability in the (M_H, m_t) plane. (From Ref. [154].)

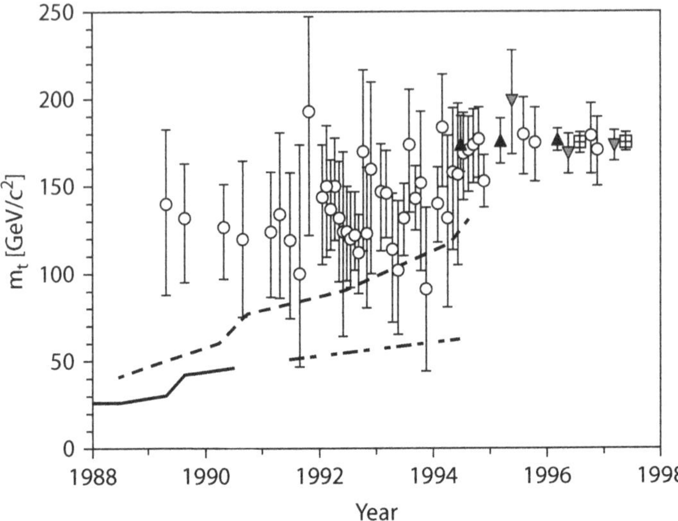

Figure 13.3 Determinations of m_t from 68% CL fits to EW observables (open circles), 95% CL lower bounds from direct searches in e^+e^- annihilations (solid line), $p\bar{p}$ collisions (dashed line) as well as from Γ_W in $\bar{p}p \to W, Z + X$ (dot-dashed line) alongside the direct measurements by CDF (triangles) and D0 (inverted triangles), as a function of the year when they were performed. (Courtesy of Chris Quigg.)

13.2 TOP QUARK DISCOVERY

Particle physicists had been convinced that the top quark must exist ever since 1977, when its partner, the bottom (or b) quark, was discovered. Little did they know it would be nearly two decades before the top was finally found. In 1985, when the FNAL TeVatron collider was first activated, the search for the top quark was well underway, while early efforts at SLAC and DESY in Germany proved fruitless. As the 1980s drew to a close, CERN, at that time running the most powerful accelerator with energies up to 315 GeV, had also failed to find the top quark. Experiments had determined that the mass of the top could be no lower than 77 GeV, well beyond the limits of CERN energy beams at the time.

In the 1990s, the focus shifted to FNAL and its two main experiments at the TeVatron: the CDF and D0 Collaborations (see Fig. 13.4 for their view). By the time researchers had begun taking data in 1992, the top mass limit had been pushed up to 91 GeV. Over the course of a decade, both the CDF and D0 Collaborations constructed enormous, complicated instruments in order to isolate the top signature.

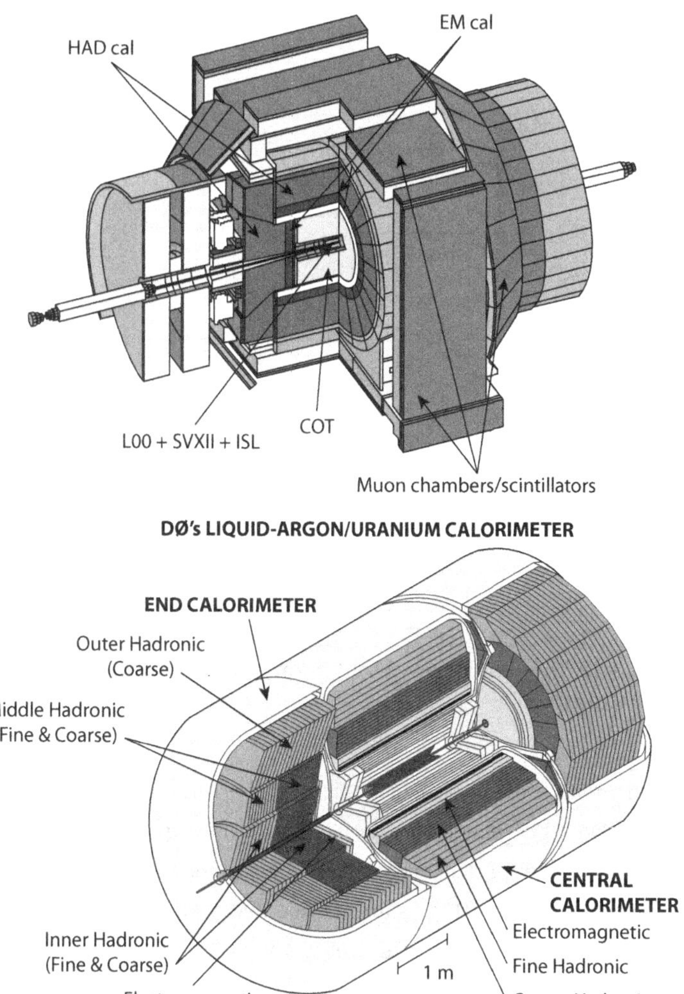

Figure 13.4 Cutaway view of the CDF (Top) and D0 (Bottom) detectors at the TeVatron.

After intensive analysis and scrutiny, the final results, made public nearly a year after researchers announced evidence for the top quark detection in April 1994, showed overwhelming evidence for it from both CDF and D0. In simultaneous publications in April 1995 [23,24], both teams reported a probability of less than one in 500,000 that their top quark candidates could be explained by background alone. The extremely large mass of the top (the current value is 172.9 GeV, similar to the mass of a gold nucleus, which contains 197

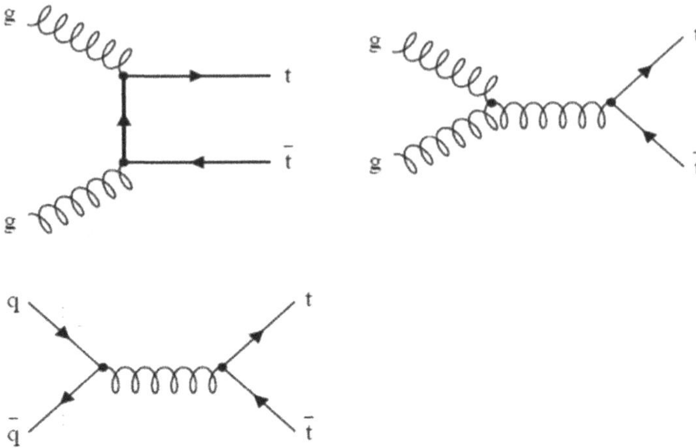

Figure 13.5 Representative Feynman diagrams leading to top quark pair production at hadron colliders: gluon-gluon fusion (top) and quark-antiquark annihilation (bottom).

protons and neutrons) suggests that it may be fundamentally different from the other quarks. The sheer enormousness of the top mass makes its decays fertile ground for new particle searches. The top also has the shortest lifetime among quarks, less than 10^{-24} s and hence decays as a free particle, the only quark to do so. All other quarks created in a collision live long enough to extract more quarks from the vacuum and (as discussed) make complicated jets composed of many particles. This top quark property has enabled the FNAL teams to determine its mass to far greater precision than the mass of any other quark.

The way the top quark was produced at the TeVatron is via the two diagrams in Fig. 13.5, *i.e.*, in pairs from QCD interactions via the processes $gg \to t\bar{t}$ (gluon-gluon fusion) and $q\bar{q} \to t\bar{t}$ (quark-antiquark annihilation), the latter dominating the former[1]. As mentioned, being the top (anti)quarks unstable, they would decay via, *e.g.*, $t\bar{t} \to (bW^+)(\bar{b}W^-) \to (bjj)(\bar{b}\ell\nu_\ell)$ + c.c., where j = jet and $\ell = e, \mu$. This is referred to as the semi-leptonic (or

[1]Notice that at the LHC it is the other way around.

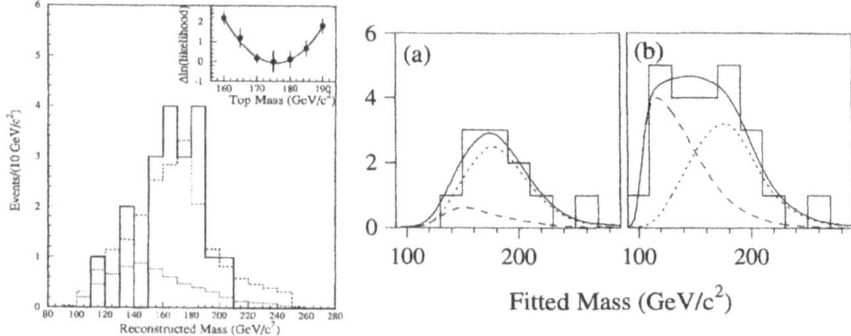

Figure 13.6 The reconstructed mass distribution of single lepton candidate events showing the data (solid histogram), the expected background and expected $t\bar{t}$ signal for the CDF (left) and D0 (right) measurements. The CDF panel inset shows the log-likelihood distribution as a function of assumed mass and the fit for the best mass value. (From Refs. [23, 24].)

semi-hadronic) decay channel of the top-antitop pair and has become over the years the preferred means to isolate this particular signal, as it combines the ability of reconstructing both heavy quark masses (as the only missing neutrino can be reconstructed by imposing $M_{\ell\nu} = M_W$) with an efficient QCD background rejection. However, in the FNAL discoveries, also the fully hadronic mode (yielding a six-jet final state, also suffering from combinatorics other than pure QCD background) and leptonic one (where two neutrino escaping detection prevent efficient reconstruction of the top quark masses) were used to establish the signal.

Fig. 13.6 shows the fitted mass distributions from the discovery papers. CDF obtained a mass of $m_t = 176 \pm 13$ GeV and D0 found $m_t = 199 \pm 30$ GeV. Using the observed yields and accounting for experimental efficiencies and acceptances, the cross section for $t\bar{t}$ pair production could be obtained: CDF found $\sigma(t\bar{t}) = 6.8^{+3.6}_{-2.4}$ pb and D0 obtained $\sigma(t\bar{t}) = 6.4 \pm 2.2$ pb. The results were consistent with each other and with the modern measurements of the mass and cross section. D0 also presented the two-dimensional plot of the mass of two jets (from hadronic W decay) versus the three jet mass (the hadronic top decay) that supported the hypothesis for the decay $t \to bW^+$. Both collaborations saw an excess which was a little less than a 5σ deviation from a background-only hypothesis, but the joint result had more than 5σ significance and originated the modern standard of requiring 5σ for a

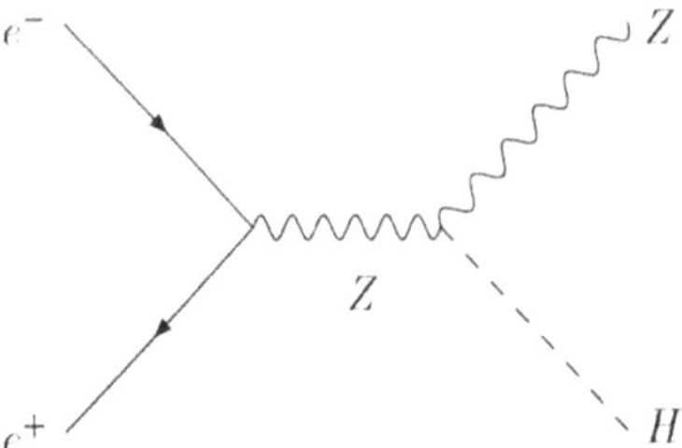

Figure 13.7 Feynman diagram for Higgs boson production in association with a Z boson in e^+e^- annihilations.

discovery. These are the kind of measurements which produced the horizontal band in Fig. 13.2.

13.3 HIGGS BOSON DISCOVERY

Another reason to deploy the LEP2 run of the CERN machine, in addition to study the properties of the $W^\pm$ boson, was the possibility of searching directly for the Higgs boson. In fact, in e^+e^- annihilation, the SM predicts the existence of the diagram in Fig. 13.7, called the Higgs-strahlung or Bjorken channel. Given the value of M_Z and the fact that the maximum energy achievable by LEP2 was eventually 209 GeV, sensitivity to M_H existed up to 119 GeV or so. As seen in Chapter 9, a Higgs boson with mass in this range would have predominantly decayed into $b\bar{b}$ pairs, so that many searches were carried out throughout the lifetime of LEP2 in any final state $HZ \to b\bar{b} + X$, where X signifies any possible decay channel of the Z boson. All of these where searched for, not just the visible lepton and quark ones, but also the case $Z \to \nu\nu$.

Near the end of the scheduled run time (*i.e.*, from the very end of the nineties), data suggested tantalising but inconclusive hints that a Higgs particle of a mass around 115 GeV might have been observed. An example of these can be seen in Fig. 13.8, where most of the excess events (above the full four-fermion background) came from 4-jet final states, $e^+e^- \to Z(\to q\bar{q})H(\to b\bar{b})$, though candidate events also existed from final states with leptons and neutrinos. As the significance was insufficient to claim a Higgs boson observation, the run time of LEP2 was extended for a few months, but to no avail, as recorded by the LEPHIGGSWG [157]. The strength of the signal remained at

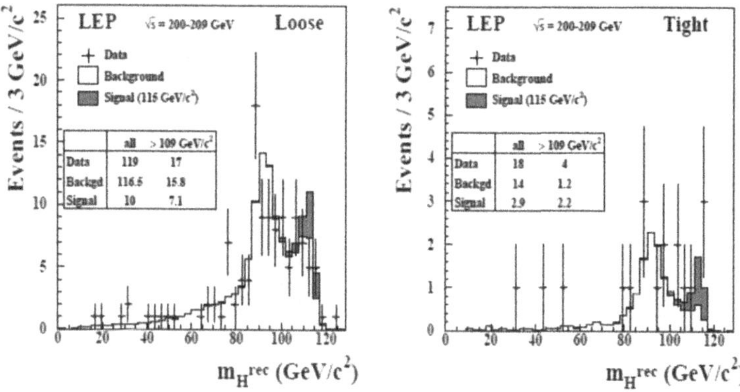

Figure 13.8 Candidate events collected by the LEP Collaborations between $\sqrt{s} = 200$ and 209 GeV in all possible HZ final states for a Higgs boson signal with $M_H = 115$ GeV using a loose (left) and tight (right) selection. (Adapted from [157].)

1.7σ; hence much less than what is required to claim a discovery (5σ). Furthermore, these events were at the extreme upper edge of the detection range of the LEP experiments, thus casting doubts on their functionality therein. There was a proposal to extend LEP operations by another year in order to seek confirmation; however, this would have delayed the start of the LHC. Finally, the decision was made to shut down LEP and progress with the LHC as planned. Therefore, for years, this observation was the only hint of a Higgs boson, which was however sufficient to start a race at the TeVatron to confirm evidence of such a particle which lasted until September 2011, when machine operation stopped: by that time, neither CDF nor D0 were able to confirm or refute these hints [159]. Yet, there is a further twist to this story, that we tell later to our patient reader.

If such excess events were instead interpreted as a background fluctuation, this enabled one to set the aforementioned 95% CL limit of $M_H = 114$ GeV, corresponding to the vertical band in Fig. 13.2. When the top quark mass had become accurately measured at TeVatron, it was possible to perform another fit, specifically, the one associated with Fig. 13.9 (left), with the $\Delta\chi^2(M_H) = \chi^2_{\rm min}(M_H) - \chi^2_{\rm min}$ curve described in Ref. [154]. The effect of the theoretical uncertainties in the SM calculations due to missing higher order corrections is shown by the thickness of the shaded curve. Including these errors, the

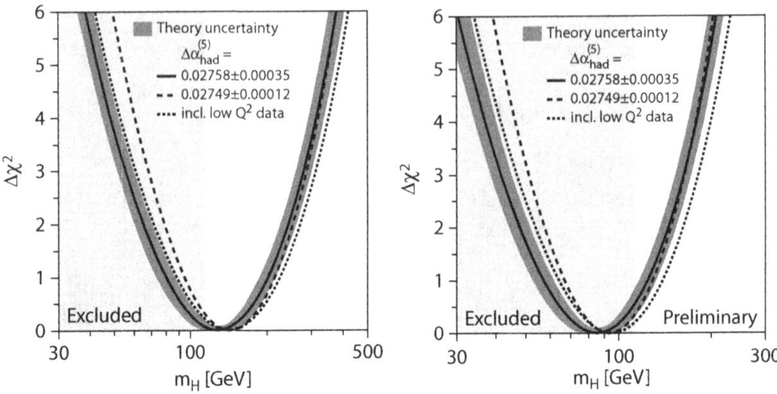

Figure 13.9 $\Delta\chi^2(M_H) = \chi^2_{\min}(M_H) - \chi^2_{\min}$ as a function of M_H as obtained in 2005 (left) and 2006 (right).

one-sided 95% CL upper limit on $\log_{10}(M_H/\text{GeV})$, given at $\Delta\chi^2 = 2.7$, is:

$$\log_{10}(M_H/\text{GeV}) \ < \ 2.455 \qquad \text{or} \qquad M_H \ < \ 285 \text{ GeV}, \tag{13.2}$$

assuming a prior probability density flat in $\log_{10}(M_H/\text{GeV})$. In case the theory driven $\Delta\alpha^{(5)}_{\text{had}}(M_Z^2)$ uncertainties (on the EM coupling constant at $\sqrt{s} = M_Z$) is allowed to vary in the range 0.02804 ± 0.00065, the central value of M_H increases while the uncertainty on M_H decreases so that the upper limit on it changes only slightly. The best fit value (minimum of the χ^2) is found at $M_H = 129\pm^{74}_{49}$ GeV, in remarkable good agreement with the value eventually measured at the LHC following the discovery of the Higgs boson.

Such a fit was done with $m_t = 178.5 \pm 3.9$ GeV, though, and the whole procedure is very sensitive to the value adopted for the top mass. Do recall that its measurements from the TeVatron were constantly evolving with time, as more and more data were being analysed by CDF and D0. So that, over the years, the outlook did not always appear as comforting as the one corresponding to the fit of [154]. For example, in Summer 2006, somewhat of an embarrassment occurred for the SM, as seen in Fig. 13.9 (right), revealing that the same fit as above ($\Delta\chi^2 = 2.7$) yielded a 95% CL upper limit on M_H of 166 GeV and, in particular, a M_H best value at 85 GeV, hence, below the direct limit from LEP2 searches (obtained with $m_t = 171.4 \pm 2.1$ GeV). Eventually, though, in the run up to the LHC, even if the final value of m_t as obtained

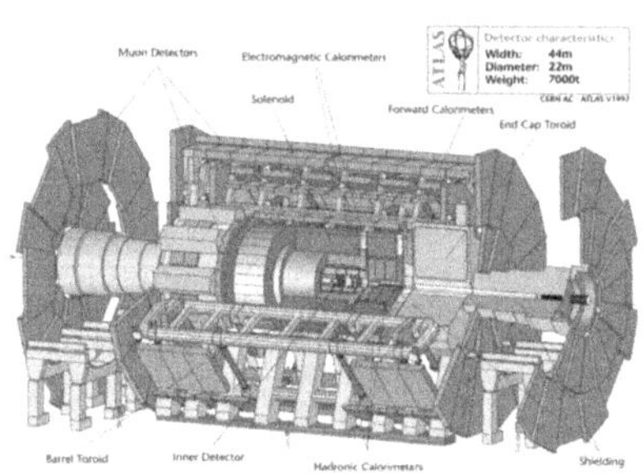

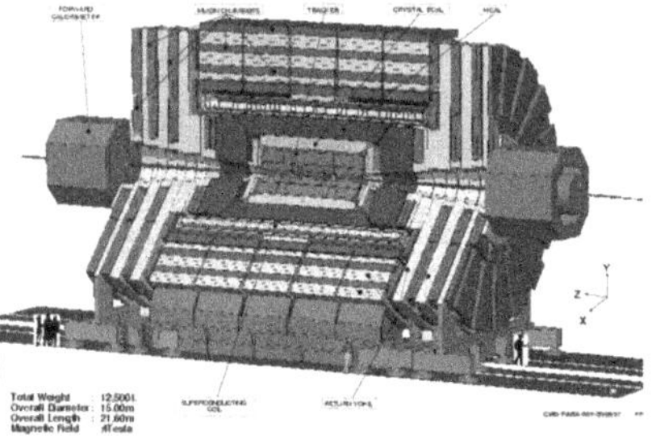

Figure 13.10 Cutaway view of the ATLAS (left) and CMS (right) detectors at the LHC.

at TeVatron continued to change, the situation did not change dramatically from the results given in Ref. [154] (*i.e.*, in Eq. (13.2)).

The Higgs boson discovery was finally announced by the ATLAS [160] and CMS [161] Collaborations at the LHC on 4 July 2012, just two days after the CDF and D0 Collaboration at the TeVatron posted their final results on their Higgs boson search [162], which revealed hints of the same particle but not enough significance to claim discovery. Evidence in the LHC detectors (see Fig. 13.10) for the new particle was presented in the three decay modes $H \to \gamma\gamma$, $H \to ZZ^* \to 4\ell$ and $H \to WW^* \to 2\ell 2\nu$ in both experiments. The combined measured mass was eventually found to be:

$$M_H = 125.09 \pm 0.21(\text{stat.}) \pm 0.11(\text{syst.})\ \text{GeV}. \tag{13.3}$$

The results for the channel $H \to \gamma\gamma$ are shown in Fig. 13.11, those for the $H \to ZZ^* \to 4\ell$ channel are shown in Fig. 13.12 while those for the $H \to WW^* \to 2\ell 2\nu$ channel are shown in Fig. 13.13. Since then, systematic studies of the Higgs boson hinted more and more to its nature being that predicted by the SM. On the one hand, its spin has been proven to be consistent with zero and its quantum numbers with a CP-even state. On the other hand, many of its couplings have been shown to follow the SM prediction, as well exemplified by Fig. 13.14.

On 8 October 2013 the Nobel Prize in Physics was awarded jointly to F. Englert and P.W. Higgs "for the theoretical discovery of a mechanism that contributes to our understanding of the origin of mass of subatomic particles,

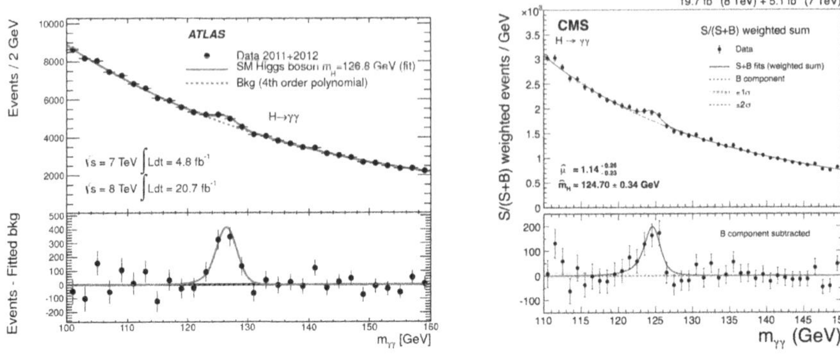

Figure 13.11 Discovery signals of the Higgs boson decaying via $H \to \gamma\gamma$ at ATLAS [160] (left) and CMS [161] (right).

and which recently was confirmed through the discovery of the predicted fundamental particle, by the ATLAS and CMS experiments at CERN's Large Hadron Collider". For the many physicists who devoted much of their academic lives, for 40 years or so, to the pursuit of the Higgs boson, it was a coronation of a dream, no less.

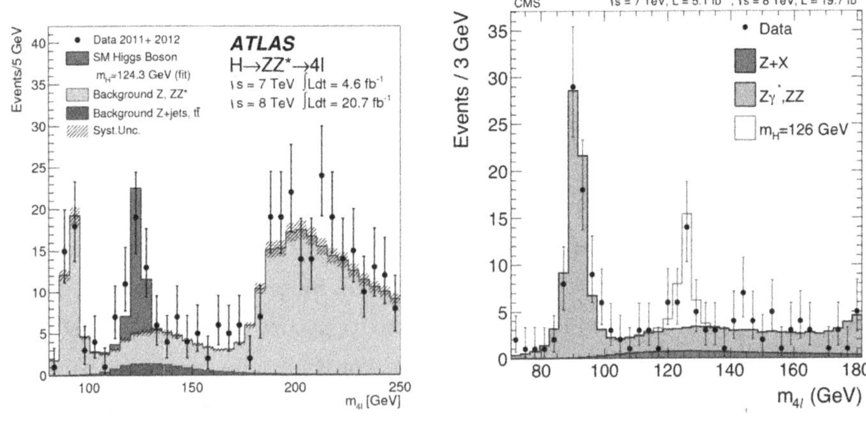

Figure 13.12 Discovery signals of the Higgs boson decaying via $H \to ZZ^* \to 4\ell$ at ATLAS [160] (left) and CMS [161] (right).

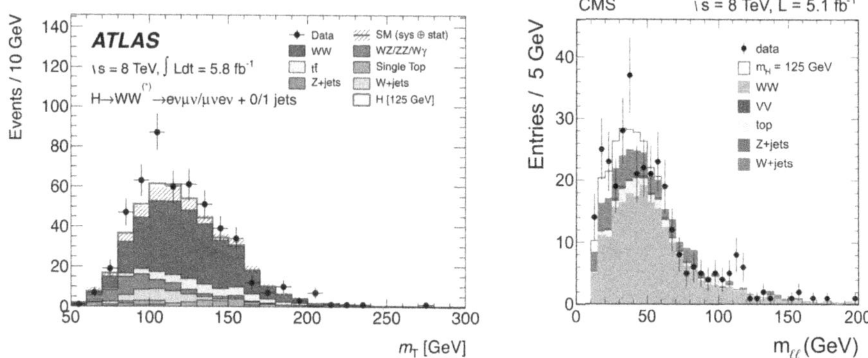

Figure 13.13 Discovery signals of the Higgs boson decaying via $H \to WW^* \to 2\ell 2\nu$ at ATLAS [160] (left) and CMS [161] (right).

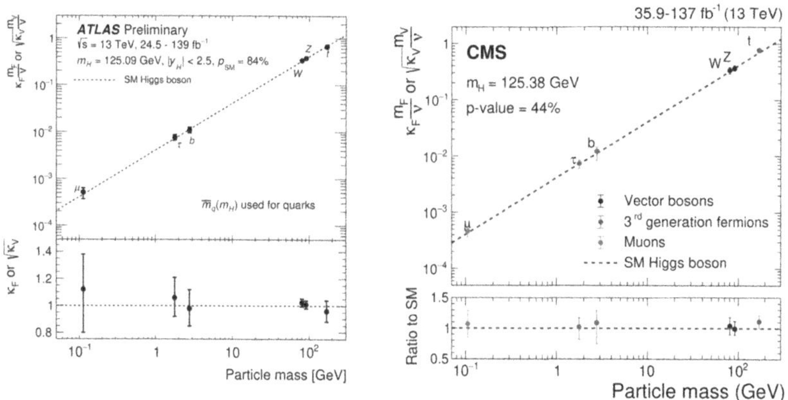

Figure 13.14 'Reduced coupling strength modifiers' $\kappa_F \frac{m_F}{v}$ for fermions ($F = t, b, \tau, \mu$) and $\sqrt{\kappa_V}\frac{m_V}{v}$ for weak gauge bosons ($V = W, Z$) as a function of their masses m_F and m_V, respectively, and the VEV of the Higgs field v, for ATLAS (left) [163] and CMS (right) [164]. The SM prediction for both cases is also shown (dotted line).

CHAPTER 14

Neutrino Masses and Mixing

As mentioned previously, there are three types of neutrinos, ν_e, ν_μ and ν_τ, which are produced along with charged leptons: e, μ and τ, respectively, in charged current weak interactions. Unlike EM charged leptons, though, neutrinos can only be detected by their weak interactions. In the SM and due to its simple structure, the neutrinos are massless. Furthermore, there are no right-handed neutrinos; hence a Dirac mass term is not allowed. Also, it has only one Higgs doublet, so a Majorana mass term is not possible either. In addition, the SM is based on a renormalisable gauge field theory, so no higher dimensional operator is permitted and therefore the Majorana mass term cannot be introduced effectively either.

While it is quite natural for neutrinos to be massless, there is no fundamental symmetry or conservation law that ensures this property. For many years, neutrinos were considered massless until recent neutrino oscillation experiments confirmed that the neutrinos have tiny masses. The first evidence for a possible neutrino oscillation was the observed deficit in the flux of electron neutrinos (about 2×10^{38} ν_e s^{-1}) produced by nuclear fusion processes in the sun [165]. It was found that the number of solar neutrinos was much less than expected. This problem was known as the solar neutrino problem. From these measurements, it was concluded that over short distances neutrinos do not change flavour; however, during long journeys neutrino flavour changes occur and this indicates that neutrinos have masses and mix. This conclusion

DOI: 10.1201/9780429443015-14

was verified by several experiments, such as Super-Kamiokande [166] and SNO [167].

14.1 NEUTRINO OSCILLATIONS

Neutrino oscillations caused by non-zero neutrino masses and mixing are a compelling solution to the solar neutrino anomaly. Furthermore, atmospheric, reactor and accelerator neutrinos experiments have provided convincing evidences for the oscillation of neutrinos. Within this oscillation scenario, neutrinos can change their flavours during their travel from the source to the detector. To understand this phenomenon explicitly and find a simple expression for the oscillation probability, we begin with the assumption of two neutrino flavours, ν_e and ν_μ. These states are called weak eigenstates and can be written as a coherent linear combination of mass eigenstates ν_1 and ν_2, which are the free particle solutions to the wave-equation and will be taken to propagate as plane waves. The weak and mass eigenstates are related by a 2×2 unitary matrix

$$\begin{pmatrix} \nu_e \\ \nu_\mu \end{pmatrix} = \begin{pmatrix} \cos\theta & \sin\theta \\ -\sin\theta & \cos\theta \end{pmatrix} \begin{pmatrix} \nu_1 \\ \nu_2 \end{pmatrix}. \tag{14.1}$$

Let us assume that at time $t = 0$ the neutrino was in a ν_e eigenstate, *i.e.*,

$$|\psi(0)\rangle = |\nu_e\rangle = \cos\theta\,|\nu_1\rangle + \sin\theta\,|\nu_2\rangle\,. \tag{14.2}$$

Then at later time t, the weak eigenstate evolves by the $e^{p_i.x}$ factor, as follows:

$$|\psi(t)\rangle = \cos\theta\,|\nu_1\rangle\,e^{-ip_1.x} + \sin\theta\,|\nu_2\rangle\,e^{-ip_2.x}, \tag{14.3}$$

where

$$p_i.x = E_i t - \vec{p_i}.\vec{x}. \tag{14.4}$$

If the neutrino travels a distance L from the source to the detector, within a time T, then the wave function at the detector is given by

$$|\psi(L,T)\rangle = \cos\theta\,|\nu_1\rangle\,e^{-i\phi_1} + \sin\theta\,|\nu_2\rangle\,e^{-i\phi_2}, \tag{14.5}$$

where $\phi_i = E_i t - p_i L$. In terms of weak eigenstates, $|\psi(L,T)\rangle$ takes the form:

$$\begin{aligned}|\psi(L,T)\rangle =& |\nu_e\rangle \left(\cos^2\theta e^{-i\phi_1} + \sin^2\theta e^{-i\phi_2}\right) \\ &+ |\nu_\mu\rangle \sin\theta\cos\theta(e^{-i\phi_2} - e^{-i\phi_1}).\end{aligned} \tag{14.6}$$

Now, it is clear that if $|\nu_1\rangle$ and $|\nu_2\rangle$ have same masses, then we get $\phi_1 = \phi_2$ and hence the second term will identically vanish. This means that the state remains $|\nu_e\rangle$ and there is no chance for any possible transition to the weak eigenstate $|\nu_\mu\rangle$. If the masses are different, then the wave function is no longer pure $|\nu_e\rangle$ and the probability of oscillation from ν_e to ν_μ is given by

$$\begin{aligned}P(\nu_e \to \nu_\mu) =& \mid \langle \nu_\mu|\psi(L,T)\rangle \mid^2 \\ =& \sin^2\theta\cos^2\theta(e^{-i\phi_2} - e^{-i\phi_1})(e^{-i\phi_2} - e^{-i\phi_1}) \\ =& \sin^2 2\theta \sin^2\left(\frac{\phi_1 - \phi_2}{2}\right).\end{aligned} \tag{14.7}$$

We will now define the phase difference term which is

$$\Delta\phi_{12} = \phi_1 - \phi_2 = (E_1 - E_2)T - \left(\mid p_1 \mid - \mid p_2 \mid\right)L. \tag{14.8}$$

One could assume $\mid p_1 \mid = \mid p_2 \mid = p$, thus

$$\begin{aligned}\Delta\phi_{12} =& \left[(p^2 + m_1^2)^{\frac{1}{2}} - (p^2 + m_2^2)^{\frac{1}{2}}\right]L, \\ \approx& \frac{m_1^2 - m_2^2}{2p}L = \frac{m_1^2 - m_2^2}{2E}L.\end{aligned} \tag{14.9}$$

In the last step, it was assumed that the neutrino is very relativistic, *i.e.*, $m \ll p$ and $L \approx cT$. Therefore, the two flavour oscillation probability is given by

$$P(\nu_e \to \nu_\mu) = \sin^2(2\theta)\sin^2\left(\frac{\Delta m_{21}^2 L}{4E}\right), \tag{14.10}$$

where $\Delta m_{21}^2 = m_2^2 - m_1^2$, while the survival probability, *i.e.*, $P(\nu_e \to \nu_e)$, is given by

$$P(\nu_e \to \nu_e) = 1 - \sin^2(2\theta)\sin^2\left(\frac{\Delta m_{21}^2 L}{4E}\right). \tag{14.11}$$

These expressions show that the oscillation probability depends on the mass squared difference. Thus, the neutrinos must be massive and their masses must also be non-degenerate in order to grant the occurrence of such oscillation.

14.2 THREE FLAVOUR NEUTRINO OSCILLATIONS

It is straightforward to extend the above analysis to three flavours of neutrinos. In this case, the weak eigenstates are ν_e, ν_μ, ν_τ and the mass eigenstates ν_1, ν_2, ν_3, which are related by a 3×3 unitary matrix

$$\begin{pmatrix} \nu_e \\ \nu_\mu \\ \nu_\tau \end{pmatrix} = \begin{pmatrix} U_{e1} & U_{e2} & U_{e3} \\ U_{\mu 1} & U_{\mu 2} & U_{\mu 3} \\ U_{\tau 1} & U_{\tau 2} & U_{\tau 3} \end{pmatrix} \begin{pmatrix} \nu_1 \\ \nu_2 \\ \nu_2 \end{pmatrix}. \tag{14.12}$$

A new matrix, called PMNS matrix [168, 169], is defined as

$$V_{\text{PMNS}} = UP, \tag{14.13}$$

where $P = \text{diag}(e^{i\rho}, e^{i\sigma}, 1)$, with ρ and σ called Majorana phases. The matrix U can be parametrised, similarly to V_{CKM}, in terms of three mixing angles and one Dirac phase. Accordingly, the pure ν_e state at time $t = 0$ is given in terms of mass eigenstates as

$$|\psi(0)\rangle = |\nu_e\rangle \equiv U_{e1}^* |\nu_1\rangle + U_{e2}^* |\nu_2\rangle + U_{e3}^* |\nu_3\rangle . \tag{14.14}$$

Then, at time t the wave function evolves to

$$|\psi(x,t)\rangle = U_{e1}^* |\nu_1\rangle e^{-i\phi_1} + U_{e2}^* |\nu_2\rangle e^{-i\phi_2} + U_{e3}^* |\nu_3\rangle e^{-i\phi_3}, \tag{14.15}$$

which can be written as

$$\begin{aligned} |\psi(x,t)\rangle = {} & (U_{e1}^* U_{e1} e^{-i\phi_1} + U_{e2}^* U_{e2} e^{-i\phi_2} + U_{e3}^* U_{e3} e^{-i\phi_3}) |\nu_e\rangle \\ & + \ (U_{e1}^* U_{\mu 1} e^{-i\phi_1} + U_{e2}^* U_{\mu 2} e^{-i\phi_2} + U_{e3}^* U_{\mu 3} e^{-i\phi_3}) |\nu_\mu\rangle \\ & + \ (U_{e1}^* U_{\tau 1} e^{-i\phi_1} + U_{e2}^* U_{\tau 2} e^{-i\phi_2} + U_{e3}^* U_{\tau 3} e^{-i\phi_3}) |\nu_\tau\rangle . \end{aligned} \tag{14.16}$$

Therefore, the oscillation probability between ν_e and ν_μ is

$$\begin{aligned} P(\nu_e \to \nu_\mu) &= |\ \langle \nu_\mu | \psi(x,t) \rangle\ |^2 \\ &= |\ U^*_{e1} U_{\mu 1} e^{-i\phi_1} + U^*_{e2} U_{\mu 2} e^{-i\phi_2} + U^*_{e3} U_{\mu 3} e^{-i\phi_3}\ |^2 . \end{aligned} \quad (14.17)$$

After some algebra, this expression can be written as

$$\begin{aligned} P(\nu_e \to \nu_\mu) =& 2\mathrm{Re}[U^*_{e1} U_{\mu 1} U_{e2} U^*_{\mu 2} [e^{i(\phi_2 - \phi_1)} - 1]] \\ &+ 2\mathrm{Re}[U^*_{e1} U_{\mu 1} U_{e3} U^*_{\mu 3} [e^{i(\phi_3 - \phi_1)} - 1]] \\ &+ 2\mathrm{Re}[U^*_{e2} U_{\mu 2} U_{e3} U^*_{\mu 3} [e^{i(\phi_3 - \phi_2)} - 1]]. \end{aligned} \quad (14.18)$$

Then, by using

$$\mathrm{Re}[e^{i(\phi_j - \phi_i)} - 1] = \cos(\phi_j - \phi_i) - 1 = -2\sin^2\left(\frac{\phi_j - \phi_i}{2}\right) = -2\sin^2\Delta_{ji}, \quad (14.19)$$

where Δ_{ji} is defined as

$$\Delta_{ji} = \frac{\phi_j - \phi_i}{2} = \left(\frac{m_j^2 - m_i^2}{4E_\nu}\right) L, \quad (14.20)$$

we find

$$\begin{aligned} P(\nu_e \to \nu_\mu) =& -4\ |\ U_{e1}\ |^2 |\ U_{e2}\ |^2 \sin^2\Delta_{21} - 4\ |\ U_{e1}\ |^2 |\ U_{e3}\ |^2 \sin^2\Delta_{31} \\ &- 4\ |\ U_{e2}\ |^2 |\ U_{e3}\ |^2 \sin^2\Delta_{32}. \end{aligned} \quad (14.21)$$

Again, the oscillation probability depends on mass squared differences, then we conclude that the neutrinos should be massive with non-degenerate masses.

14.3 NEUTRINO MASSES AND MIXING

The flavour eigenstates, ν_e, ν_μ and ν_τ, are defined as the particles produced in association with definite charged leptons: *i.e.*, electron, muon and tauon, respectively. For example, the neutrinos emitted in weak processes such as the β-decay or pion decay with electron or muon are termed the electron or muon neutrinos. In the SM, the neutrinos ν_e, ν_μ and ν_τ are described by the fields that form the weak doublets (or weak charged currents) with charged lepton fields of definite mass:

$$J^\mu = \bar{l}\gamma^\mu(1-\gamma_5)\nu_l, \quad l = e, \mu, \tau. \tag{14.22}$$

Flavour mixing means that the flavour neutrinos ν_α $(\alpha = e, \mu, \tau)$ are not the same as neutrinos of definite mass ν_i $(i = 1, 2, 3)$. The electron, muon and tauon neutrino states have no definite masses but turn out to be coherent combinations of the mass states. The weak charged current processes mix neutrino mass states. The flavour eigenstates $\nu_\alpha^T \equiv (\nu_e, \nu_\mu, \nu_\tau)$ are expressed in terms of the mass eigenstates $\nu_i^T \equiv (\nu_1, \nu_2, \nu_3)$ as

$$\nu_\alpha = \sum_i (U_{\text{PMNS}})_{\alpha i}\nu_i, \tag{14.23}$$

where U_{PMNS} is the 3×3 PMNS lepton mixing matrix. Inserting Eq. (14.23) into Eq. (14.22) we can write the weak charged currents as

$$J^\mu = \bar{l}\gamma^\mu(1-\gamma_5)U_{\text{PMNS}}\nu. \tag{14.24}$$

Thus, the lepton mixing matrix connects the neutrino mass fields and charged lepton fields with definite mass in the weak charged currents.

If one assumes that the neutrinos are Majorana fermions, the mass matrix of neutrinos can be diagonalised by one unitary matrix S_ν as

$$S_\nu^T \mathcal{M}_\nu S_\nu = M_\nu^d. \tag{14.25}$$

This diagonalisation can be achieved by rotating the flavour basis as

$$\nu' = S_\nu \nu. \tag{14.26}$$

The charged lepton mass matrix, in general, can be diagonalised by two unitary matrices:

$$S_l M_l V_l^\dagger = M_l^d. \tag{14.27}$$

This can be implemented by the following rotations:

$$l'_L = S_l l_L, \qquad l'_R = V_l l_R, \tag{14.28}$$

where $M_\nu^d \equiv \text{diag}(m_1, m_2, m_3)$ and $M_\ell^d \equiv \text{diag}(m_e, m_\mu, m_\tau)$ are the diagonal matrices. Plugging these relations into the charged current, we obtain

$$J^\mu = \bar{l}'\gamma^\mu(1-\gamma_5)\nu' = \bar{l}\gamma^\mu(1-\gamma_5)S_l^\dagger S_\nu \nu. \tag{14.29}$$

Thus, the physical neutrino mixing matrix is given by

$$U_{\rm PMNS} \;=\; S_l^\dagger S_\nu. \tag{14.30}$$

For the case of three Majorana neutrinos, we can parametrise the mixing matrix as

$$U_{\rm PMNS} = U_{23}(\theta_{23})U_{13}(\theta_{13},\delta)U_{12}(\theta_{12})I_\phi \equiv V I_\phi, \tag{14.31}$$

where the U_{ij} are matrices of rotations in the ij plane by the angle θ_{ij} while δ is the Dirac CP-violating phase attached to the 1–3 rotation. Here, $I_\phi \equiv {\rm diag}(1, e^{i\phi_1}, e^{i\phi_2})$ is the diagonal matrix of the Majorana CP-violating phases. Moreover, V can be parameterised as follows [170]:

$$V = \begin{pmatrix} c_{12}c_{13} & s_{12}c_{13} & s_{13}e^{-i\delta} \\ -s_{12}c_{23} - c_{12}s_{23}s_{13}e^{i\delta} & c_{12}c_{23} - s_{12}s_{23}s_{13}e^{i\delta} & s_{23}c_{13} \\ s_{12}s_{23} - c_{12}c_{23}s_{13}e^{i\delta} & -c_{12}s_{23} - s_{12}c_{23}s_{13}e^{i\delta} & c_{23}c_{13} \end{pmatrix}, \tag{14.32}$$

where c_{ij} and s_{ij} refer to $\cos(\theta_{ij})$ and $\sin(\theta_{ij})$, respectively.

14.4 EXPERIMENTAL RESULTS ON MASSES AND MIXING ANGLES

The values of the mixing angles and mass squared differences are measured by neutrino oscillation experiments as shown in Tab. 14.1. There is currently no constraint on any of the CP odd phases or on the sign of Δm_{13}^2. Note that, in contrast to the quark sector, we have two large angles (one possibly maximal) and one small (possibly zero) angle. From the experimental results, we find that the experiments do not give the actual masses of the three types of neutrinos but give only the mass squared differences between them.

From these results, there are three possible scenarios for the neutrino mass relations.

1. Normal hierarchy, *i.e.*, $m_1 < m_2 \ll m_3$. In this case $\Delta m_{23}^2 \equiv m_3^2 - m_2^2 > 0$ and $m_3 \simeq \sqrt{\Delta m_{23}^2} \simeq 0.03 - 0.07$ eV. The solar neutrino oscillation

Experiment	Masses	Angles
-Solar experiments: SNO [167] and Super-Kamiokande -Reactor experiment: KamLAND	$\Delta m_{12}^2 = (7.9 \pm 0.4) \times 10^{-5}\text{eV}^2$	$\theta_{12} = 33.9° \pm 1.6°$
-Atmospheric experiments: Super-Kamiokande [166] -Accelerator experiments: K2K and MINOS	$\lvert\Delta m_{32}^2\rvert = (2.4 \pm 0.3) \times 10^{-3}\ \text{eV}^2$	$\sin^2\theta_{23} = 0.47$
CHOOZ experiment [171]		$\sin^2\theta_{13} < 0.048$

Table 14.1 Neutrino masses and mixing from different neutrino experiments.

involves the two lighter levels. The mass of the lightest neutrino is unconstrained. If $m_1 \ll m_2$, then we find the value of $m_2 \simeq 0.008$ eV and $m_1 \sim 0$.

2. Inverted hierarchy, *i.e.*, $m_1 \simeq m_2 \gg m_3$ with $m_{1,2} \simeq \sqrt{\Delta m_{23}^2} \simeq 0.03-0.07$ eV. In this case, solar neutrino oscillation takes place between the heavier levels and we have $\Delta m_{23}^2 \equiv m_3^2 - m_2^2 < 0$. We have no information about m_3 except that its value is much less than the other two masses.

3. Degenerate neutrinos, *i.e.*, $m_{\nu_1} \simeq m_{\nu_2} \simeq m_{\nu_3} \simeq \tilde{m}$. Here, from the astrophysical constraint $\sum_i m_{\nu_i} < 1$ eV, one finds that $\tilde{m} < 0.3$ eV.

14.5 ORIGIN OF NEUTRINO MASS

As advocated above, in the SM, the left-handed neutrinos that form the EW doublets, ψ_{lL}, have zero electric charge and no colour. The right-handed components, ν_R, are not included. There are two ways to construct the mass term of a fermion in the Lagrangian, the first way is to mix the left and right components of the same field and this term is called the 'Dirac mass term' and the particle in this case is called Dirac particle, *e.g.*,

$$\mathcal{L}_D = \bar{\psi}_L m_D \psi_R + \bar{\psi}_R m_D \psi_L. \tag{14.33}$$

The second way is to connect the left or the right components with their conjugate fields and this mass type is called a Majorana mass and the particle in this case is called a Majorana particle, *e.g.*,

$$\mathcal{L}_M = \bar{\psi}^c_L m_M \psi_L + \bar{\psi}^c_R m'_M \psi_R, \tag{14.34}$$

where c refers to the charge-conjugation. It is important to note that the Majorana mass term breaks any global or gauge symmetry imposed on the model. As an example, if the left-handed component of the field ψ_L has an electric charge of a given value, then the conjugate field $\bar{\psi}^c_L$ has the same value of the electric charge. This implies that the Majorana mass term breaks the corresponding $U(1)_{\rm EM}$ symmetry. In other words, the particle that has a Majorana mass must be singlet under all global and gauge symmetries.

For a neutrino in the SM, we cannot construct a Dirac mass term due to the absence of ν_R. Also, the Majorana mass term breaks the lepton number symmetry by two units. Therefore, the neutrino is expected to be a massless fermion. However, several experimental evidences like the above mentioned neutrino oscillations confirmed the existence of non-vanishing neutrino masses. This gives an important hint for BSM physics.

14.6 RIGHT-HANDED NEUTRINOS AND SEESAW MECHANISMS

Now, we consider the possibility of extending the SM by adding three right-handed neutrinos (singlet under all SM gauge symmetries). In this case, a neutrino Yukawa coupling

$$\lambda_\nu \bar{L} \Phi \nu_R + h.c. \tag{14.35}$$

may exist. Hence, after EWSB, Dirac neutrino masses are obtained as

$$m_D = \lambda_\nu \langle \Phi \rangle. \tag{14.36}$$

Thus, the observed neutrino masses would require $\lambda_\nu \leq 10^{-13} - 10^{-12}$ GeV. If ν_R is the same type of field as the right-handed components of other fermions, such smallness looks rather unnatural.

14.6.1 Type I Seesaw

The right-handed neutrinos, as neutral particles, are allowed to have Majorana masses

$$M_R \bar{\nu}_R^c \nu_R + h.c. \tag{14.37}$$

In this case the neutrino mass matrix is given by

$$M_\nu = \begin{pmatrix} 0 & m_D^T, \\ m_D & M_R. \end{pmatrix}. \tag{14.38}$$

The mixing between the left- and right-handed components leads, after the diagonalisation of the mass matrix, to two different eigenstates called the light and heavy neutrinos with the following eigenvalues:

$$\mathcal{M}_\nu \simeq -m_D^T M_R^{-1} m_D, \tag{14.39}$$

$$\mathcal{M} \simeq M_R. \tag{14.40}$$

The first eigenvalue is the mass of the light neutrino while the other is the mass of the heavy neutrino. The mass of the light neutrino is of the order of $(10^{-11} - 10^{-12})$ GeV; thus a M_R of order 10^{14} GeV is assumed.

This mechanism is called *Type I seesaw* [172, 173], which provides an elegant explanation of the smallness of the neutrino masses and is usually described diagrammatically as in Fig. 14.1.

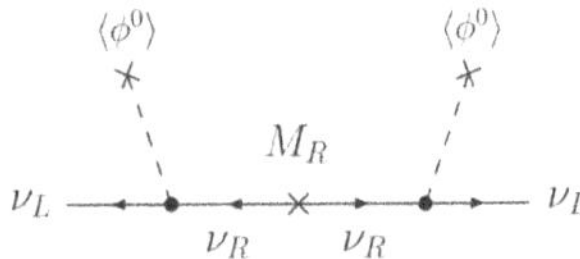

Figure 14.1 Type I seesaw realisation of the small Majorana mass for the left-handed neutrino.

14.6.2 Type II Seesaw

Instead of extending the SM by adding heavy singlet fermions as in the previous section, one can make use of the fact that $\overline{\ell_L}\,\ell_L^c$ is an $SU(2)_L$ triplet and introduce a heavy *triplet scalar* in the Higgs sector [174–176], so that a gauge invariant and renormalisable $\overline{\ell_L}\,\ell_L^c$-type mass term can be formed. Specifically, suppose we have a heavy $SU(2)_L$ triplet scalar field Δ with hypercharge $Y = -2$ and a convenient 2×2 matrix parameterisation given by

$$\Delta = \begin{pmatrix} \Delta^-/\sqrt{2} & \Delta^{--} \\ \Delta^0 & -\Delta^-/\sqrt{2} \end{pmatrix}, \tag{14.41}$$

then the Lagrangian

$$-\mathcal{L}_{\text{Type II}} = \frac{Y_\Delta}{2}\,\overline{\ell_L} i\tau_2\,\Delta\,\ell_L^c + \mu_\Delta\,\phi^T\Delta\,\phi + M_\Delta^2\,\Delta^\dagger\Delta + h.c. \tag{14.42}$$

will give rise to the process depicted in Fig. 14.2a. This then leads to an effective mass term

$$\begin{aligned} m_{\text{eff}}^{\text{II}} &\simeq \mu_\Delta\,Y_\Delta\frac{\langle\phi^0\rangle^2}{M_\Delta^2}, \\ &= \lambda_\Delta\,Y_\Delta\frac{\langle\phi^0\rangle^2}{M_\Delta} \quad \text{(after setting} \quad \mu_\Delta \equiv \lambda_\Delta M_\Delta). \end{aligned} \tag{14.43}$$

This expression has the same form as Eq. (14.39) with the couplings $\lambda_\Delta Y_\Delta$ playing the role of y_ν^2. So, when $M_\Delta \gg \langle\phi^0\rangle$, small neutrino masses can be induced. This mechanism is known as *Type II seesaw* [174].

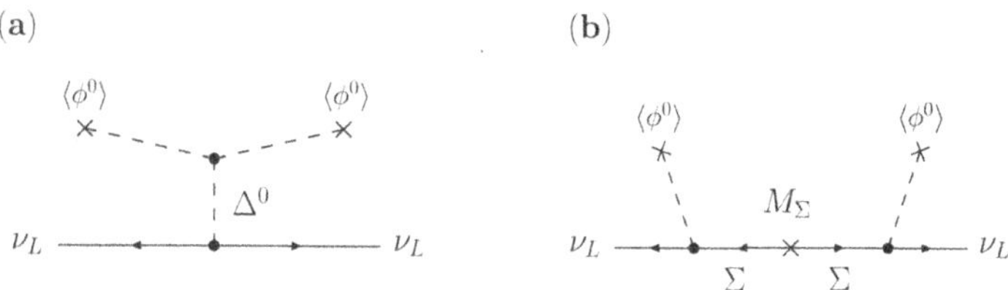

Figure 14.2 **(a)** The process induced by the Type II seesaw Lagrangian that will give rise to small neutrino Majorana masses. **(b)** The corresponding process in the Type III seesaw case with heavy triplet fermion Σ instead.

14.6.3 Type III Seesaw

Another possibility is to replace the right-handed neutrinos with heavy triplet fermions and allow them to interact with the ordinary lepton doublets via Yukawa couplings [177]. In this scenario, the Higgs sector is unmodified and a set of self-conjugate $SU(2)_L$ triplets of exotic leptons with hypercharge $Y = 0$ are added:

$$\Sigma = \begin{pmatrix} \Sigma^- & \Sigma^0/\sqrt{2} \\ \Sigma^0/\sqrt{2} & \Sigma^+ \end{pmatrix}. \tag{14.44}$$

The corresponding Lagrangian for this model is given by

$$-\mathcal{L}_{\text{Type III}} = Y_\Sigma \, \overline{\ell_L} \, i\tau_2 \, \Sigma \, \phi + M_\Sigma \, \text{Tr} \left(\overline{\Sigma^c} \, \Sigma \right) + h.c. \tag{14.45}$$

This gives rise to the diagram shown in Fig. 14.2b and, after integrating out the heavy Σ field, one obtains the desired form for the seesaw neutrino mass

$$m_{\text{eff}}^{\text{III}} \simeq Y_\Sigma \, \frac{\langle \phi^0 \rangle^2}{M_\Sigma} \, Y_\Sigma^T \, . \tag{14.46}$$

Hence, by setting $M_\Sigma \gg \langle \phi^0 \rangle$, one can explain the smallness of neutrino masses and, as a result, this is often referred to as the *Type III seesaw* mechanism [178].

14.6.4 Inverse Seesaw

In the Type I seesaw mechanism, we assume that the M_R is a very large scale. From another point view, M_R represents the strength of breaking the lepton number in the Lagrangian. That is, in the seesaw Type I we assume that the lepton number is strongly violated, which is not a widely accepted assumption in the particle physics community. As a result, another types of seesaw mechanisms that respects[1] the lepton number arose. One of these mechanisms is the so called *Inverse Seesaw* (or IS) mechanism [179]. The construction of this mechanism requires the existence of another SM singlet fermions S other than the right-handed neutrino ν_R. Explicitly, we have to add a fermion S whose lepton number is opposite to that of ν_R. In this case,

[1] Here, 'respects' is used in the sense that the violation of the lepton number must be mild.

the most general neutrino part of the Lagrangian is given by

$$\begin{aligned}\mathcal{L}_{\rm IS}^{\nu} &= y_\nu \overline{\ell_L}\tilde{\phi}\nu_R + M_R\overline{(\nu_R)^c}S + y_S\overline{\ell_L}\tilde{\phi}S \\ &+ \mu_R\overline{(\nu_R)^c}\nu_R + \mu_S\overline{(S)^c}S + h.c.,\end{aligned} \tag{14.47}$$

where y_S, μ_R and μ_S are naturally small because of the so called 't Hooft criteria [180]. Indeed, in the limit y_S, μ_R, $\mu_S \rightarrow 0$, the lepton number is restored as a conserved symmetry.

After EWSB, the mass matrix of the neutrinos is given, for one generation, by

$$\begin{pmatrix} 0 & m_D & M_S \\ m_D^T & \mu_R & M_R \\ M_S^T & M_R^T & \mu_S \end{pmatrix}, \tag{14.48}$$

where $m_D = y_\nu\, v$ and $M_S = y_S\, v$. As mentioned before, $M_S, \mu_R, \mu_S \ll m_D, M_R$; thus the neutrino masses can be given, with a very good approximation, by

$$m_{\nu_\ell} = \frac{m_D(m_D\,\mu_S - 2M_R M_S)}{M_R^2 + m_D^2}, \tag{14.49}$$

$$\begin{aligned} m_{\nu_{H,H'}} &= \frac{M_R(M_R\,\mu_S + 2m_D M_S)}{2(M_R^2 + m_D^2)} + \frac{\mu_R}{2} \\ &\mp \sqrt{M_R^2 + m_D^2}. \end{aligned} \tag{14.50}$$

It is worth mentioning that, in the IS scenario, the neutrino Yukawa coupling can be of order $\mathcal{O}(1)$ and the large scale M_R can be brought to the TeV scale. This is because the suppression factors needed to account for light neutrino masses are played by the naturally small parameters M_S and μ_S instead of the Yukawa coupling y_ν. Indeed, if $y_\nu \sim \mathcal{O}(1)$, $M_R \sim 1$ TeV and $\mu_S \sim 10\ M_S \sim \mathcal{O}(10^{-7})$ GeV, then a 1 eV neutrino mass can be obtained.

Epilogue

We hope to have given the reader a brief, yet comprehensive, overview of the SM. We have exalted its theoretical elegance and pinpointed a few key experimental discoveries from the past four decades (and related measurements) that confirm it beautifully. This applies to all its components, *i.e.*, the matter sector (with the discovery of the top quark in 1995), the gauge structure (with the discoveries of the $W^{\pm}$ and Z bosons in 1983) as well as the most intriguing of all its aspects, the one connected with mass generation (with the discovery of the Higgs boson in 2012). This has been the triumph of particle physics at the EW scale.

However, while these discoveries may have seemed as the last steps in the long journey to describe the particle physics world, we cannot refrain in ending this book from noticing that the SM is unquestionably incomplete. There are in fact other experimental discoveries that cannot be reconciled with it. First, we should recall here the fact that DM accounts for 23% of the mass/energy density of the observable Universe, while the only plausible candidate from the SM, the neutrino, provides too modest a level of it. Second, in the SM, neutrinos are massless by construction, yet recent experiments indicate that this is not true, as their (leptonic) flavour can oscillate, which in turn implies that neutrinos do have (small) masses. Third, the SM cannot explain the observational fact that the Universe is made of matter and not antimatter, as it is now firmly established that the strength of CP violation in it, which is the essential ingredient for explaining the cosmological baryon asymmetry, is not sufficient.

Furthermore, from a theoretical point of view, it is also clear that the SM cannot be a valid realisation of the ultimate theory of matter and forces, valid to high energies (*e.g.*, up to the Planck scale, where gravity becomes strong and should unify with the other interactions). In fact, there are theoretical inconsistencies leading to instabilities of the Higgs boson mass that systematically grow with the energy at which the SM is being probed, ultimately leading to either its breakdown or a solution perceived as highly *ad hoc*, hence unnatural. This conceptual flaw is known as the 'hierarchy problem' and it is associated with the absence of a symmetry protecting the Higgs mass, which is necessarily generated at the EW scale, when the interactions of the Higgs field are instead generated where the natural cut-off scale is or indeed above it, *i.e.*, the energy at which a GUT (eventually also including gravity) will inevitably have to be formulated, $M_{\rm GUT} \simeq 3 \times 10^{16}$ GeV. This phenomenon is all the more puzzling if one realises that, in essence, this means that the Higgs mechanism, designed to explain mass, is somehow unable to cope with the presence of the force which exists solely because of the very same existence of mass, gravity.

We are intimately sure that this sort of, so-to-say, *inner conflict* within the SM itself will actually be the key to unlock a new Pandora's box of discoveries. What has been the confirmation of the cornerstone of the SM, a Higgs boson giving mass to itself and all objects, should now turn out to be the stepping stone into a new physics world. The latter is currently unknown to us. We know though what its dynamics should prevent, *i.e.*, the aforementioned hierarchy problem of the SM. We also know what it should generate, *i.e.*, a mass for neutrinos. All this should be interlocked with the existence of DM and the presence of additional CP violation.

We have searched for a new theory offering all this already, in fact, for as many decades as it took to confirm the SM, and nothing has ever been found that an experiment could confirm. This should not deter us in any way, though. There is value in this 'Negative Knowledge'[1]. We uphold the words of [181]: "Building on existing conceptions of negative knowledge, we systematically relate the concept to research on expertise and learning from errors". Perhaps, our biggest mistake, as a community of particle physicists, has been that of searching for new physics beyond the SM before actually confirming the latter in all its foundations. We have embraced beautiful theories which would solve at once all of the above problems of the SM, like Supersymmetry for one. We have written another whole book [182] about its beauty and simplicity, somehow invoking its inevitability. Were we wrong? That is still to be decided

[1]In terms of its theoretical foundations, the concept relates to constructivist theorisation and metacognition; however, in practice, negative knowledge is nothing but experimentally acquired knowledge about what is wrong, which is most valuable per se.

(by what future experimental data will hold for us). However, we acknowledge now that we had perhaps forgotten the warning of Sir K.R. Popper: "Whenever a theory appears to you as the only possible one, take this as a sign that you have neither understood the theory nor the problem which it was intended to solve" (from *Objective Knowledge: An Evolutionary Approach, 1972*). Now we have learnt our lesson, which is what has brought us back to the SM. But no, we do not believe that there is a desert ahead of us! The EW and Planck scales are closer that one may think now. There will be a bridge between them too.

APPENDIX A

Feynman Rules

The Feynman rules presented in this appendix have been generated by SARAH 4, a model builder's tool [183]. Here, $\xi_{G,A,W,Z}$ refers to the general gauge parameter for a g, γ, $W^\pm, Z$ boson, respectively. Furthermore, $\eta_{G,\gamma,\pm,Z}$ refers to the ghost associated with a $g, \gamma, W^\pm, Z$ state, respectively. Then, $G^{\pm,0}$ refers to the Goldstone associated with a $W^\pm, Z$ state, respectively. (For the usual expressions of the external fermion spinors and gauge boson polarisation vectors we refer to, *e.g.*, [42].) Also notice that momenta appearing in Feynman rules are intended as entering into the vertices. Finally, hereafter, we use the following definitions for the $g_{1,2,3}$ couplings of $U(1)_Y, SU(2)_L$ and $SU(3)_C$, respectively: $g_1 = e\cos\theta_W$, $g_2 = e\sin\theta_W$ and $g_3 = g_s$.

DOI: 10.1201/9780429443015-A

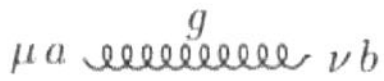

μa (g) νb $$-i\delta_{ab}\left[\frac{g_{\mu\nu}}{p^2+i\epsilon}-(1-\xi_G)\frac{p_\mu p_\nu}{(p^2)^2}\right] \tag{A.1}$$

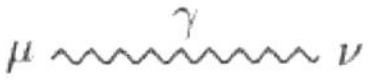

μ (γ) ν $$-i\left[\frac{g_{\mu\nu}}{p^2+i\epsilon}-(1-\xi_A)\frac{p_\mu p_\nu}{(p^2)^2}\right] \tag{A.2}$$

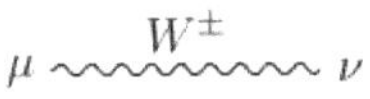

μ ($W^\pm$) ν $$-i\,\frac{1}{p^2-M_W^2+i\epsilon}\left[g_{\mu\nu}-(1-\xi_W)\frac{p_\mu p_\nu}{p^2-\xi_W M_W^2}\right] \tag{A.3}$$

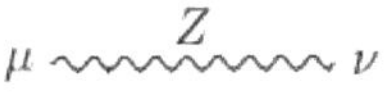

μ (Z) ν $$-i\,\frac{1}{p^2-M_Z^2+i\epsilon}\left[g_{\mu\nu}-(1-\xi_Z)\frac{p_\mu p_\nu}{p^2-\xi_Z M_Z^2}\right] \tag{A.4}$$

p $$\frac{i(\not{p}+m_f)}{p^2-m_f^2+i\epsilon} \tag{A.5}$$

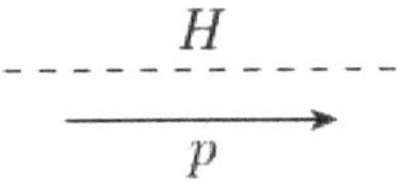

H, p $$\frac{i}{p^2-m_H^2+i\epsilon} \tag{A.6}$$

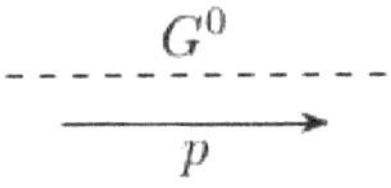

G^0, p $$\frac{i}{p^2-\xi_Z M_Z^2+i\epsilon} \tag{A.7}$$

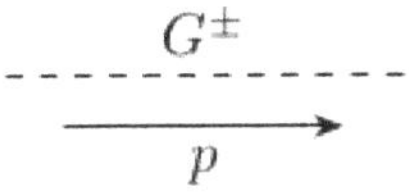

$G^\pm$, p $$\frac{i}{p^2-\xi_W M_W^2+i\epsilon} \tag{A.8}$$

Vertices	Coupling	
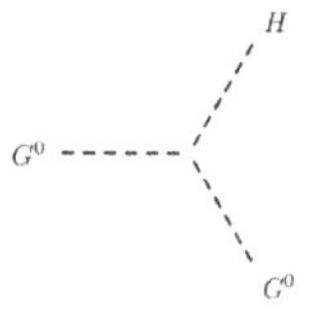 G^0 H G^0	$-iv\lambda$	(A.9)
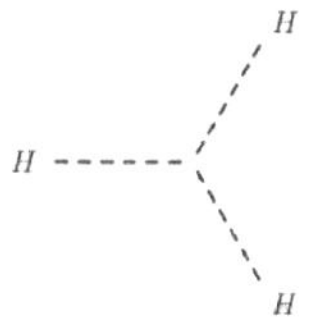 H H H	$-3iv\lambda$	(A.10)
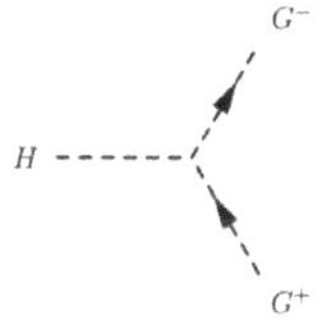 H G^- G^+	$-iv\lambda$	(A.11)
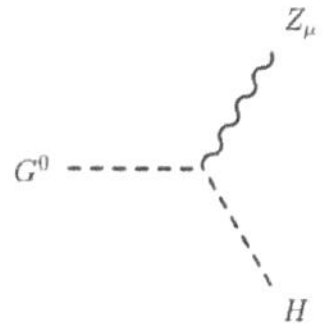 G^0 Z_μ H	$\frac{1}{2}\Big(- g_1 \sin\theta_W - g_2 \cos\theta_W\Big)\Big(- p_\mu^H + p_\mu^{G^0}\Big)$	(A.12)
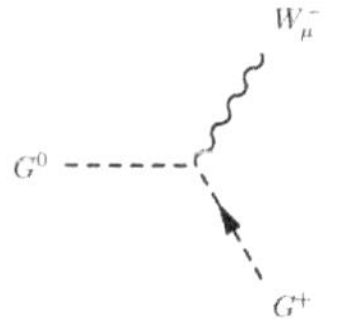 G^0 W_μ^- G^+	$\frac{1}{2}g_2\Big(- p_\mu^{G^+} + p_\mu^{G^0}\Big)$	(A.13)
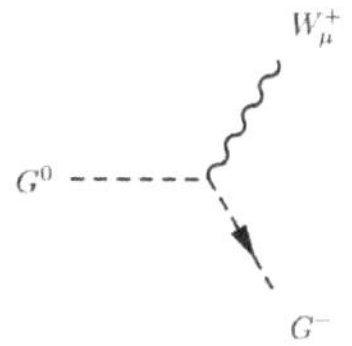 G^0 W_μ^+ G^-	$\frac{1}{2}g_2\Big(- p_\mu^{G^-} + p_\mu^{G^+}\Big)$	(A.14)

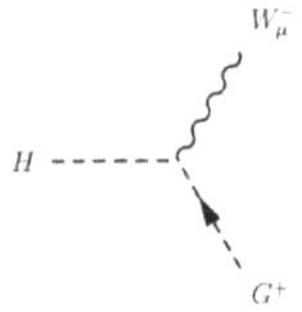

$$\frac{i}{2}g_2\Big(-p_\mu^{G^+}+p_\mu^H\Big) \tag{A.15}$$

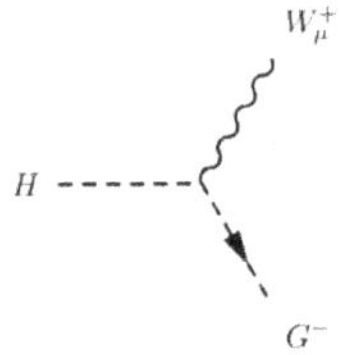

$$-\frac{i}{2}g_2\Big(-p_\mu^{G^-}+p_\mu^H\Big) \tag{A.16}$$

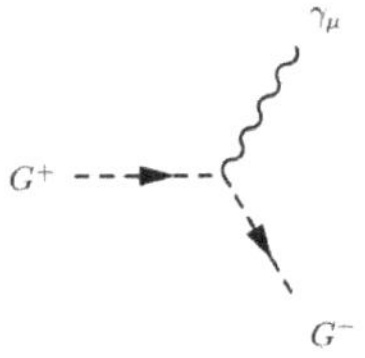

$$-\frac{i}{2}\Big(g_1\cos\theta_W+g_2\sin\theta_W\Big)\Big(-p_\mu^{G^-}+p_\mu^{G^+}\Big) \tag{A.17}$$

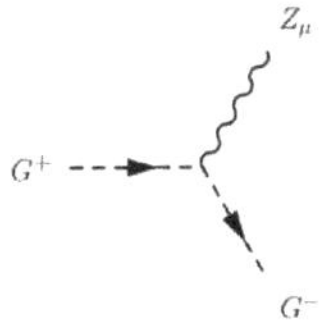

$$-\frac{i}{2}\Big(-g_1\sin\theta_W+g_2\cos\theta_W\Big)\Big(-p_\mu^{G^-}+p_\mu^{G^-}\Big) \tag{A.18}$$

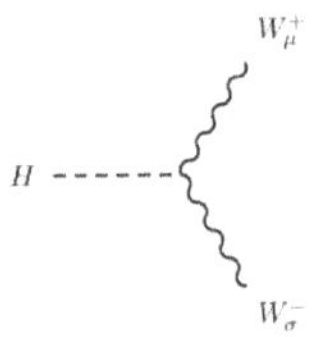

$$\frac{i}{2}g_2^2v\Big(g_{\sigma\mu}\Big) \tag{A.19}$$

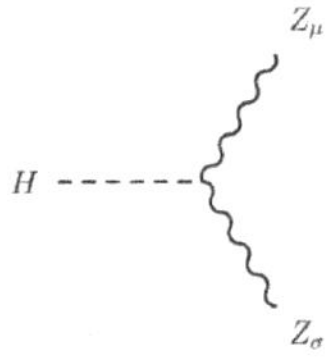

$$\frac{i}{2}v\Big(g_1\sin\theta_W+g_2\cos\theta_W\Big)^2\Big(g_{\sigma\mu}\Big) \tag{A.20}$$

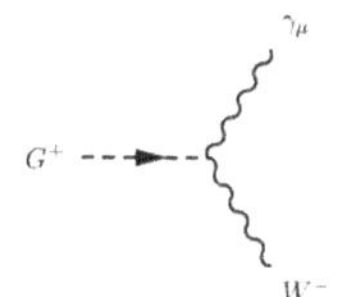

$$\frac{i}{2} g_1 g_2 v \cos\theta_W \Big(g_{\sigma\mu} \Big) \tag{A.21}$$

$$-\frac{i}{2} g_1 g_2 v \sin\theta_W \Big(g_{\sigma\mu} \Big) \tag{A.22}$$

$$\frac{i}{2} g_1 g_2 v \cos\theta_W \Big(g_{\sigma\mu} \Big) \tag{A.23}$$

$$-\frac{i}{2} g_1 g_2 v \sin\theta_W \Big(g_{\sigma\mu} \Big) \tag{A.24}$$

$$\begin{aligned} &-\tfrac{i}{2} g_3 \delta_{ij} \lambda^c_{a,b} \Big(\gamma_\mu \cdot \tfrac{1-\gamma_5}{2} \Big) \\ &-\tfrac{i}{2} g_3 \delta_{ij} \lambda^c_{a,b} \Big(\gamma_\mu \cdot \tfrac{1+\gamma_5}{2} \Big) \end{aligned} \tag{A.25}$$

$$\begin{aligned} &-\tfrac{i}{6} \delta_{ab} \delta_{ij} \Big(-3 g_2 \sin\theta_W + g_1 \cos\theta_W \Big) \Big(\gamma_\mu \cdot \tfrac{1-\gamma_5}{2} \Big) \\ &+ \tfrac{i}{3} g_1 \cos\theta_W \delta_{ab} \delta_{ij} \Big(\gamma_\mu \cdot \tfrac{1+\gamma_5}{2} \Big) \end{aligned} \tag{A.26}$$

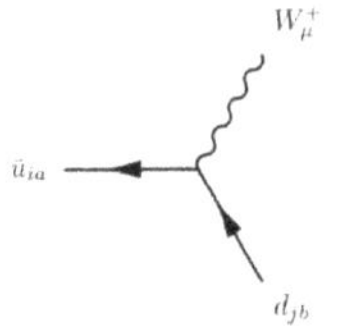

$\bar{u}_{ia}$, d_{jb}, W^+_μ

$$-i\frac{1}{\sqrt{2}}g_2\delta_{ab}\sum_{a=1}^{3}V^{d,*}_{L,ja}V^{u}_{L,ia}\Big(\gamma_\mu\cdot\frac{1-\gamma_5}{2}\Big) \quad (A.27)$$

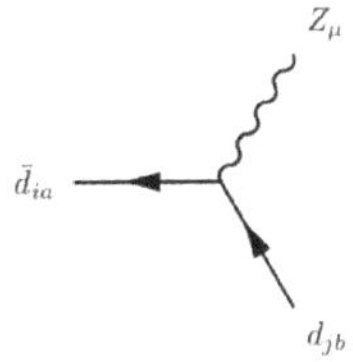

$\bar{d}_{ia}$, d_{jb}, Z_μ

$$\frac{i}{6}\delta_{ab}\delta_{ij}\Big(3g_2\cos\theta_W+g_1\sin\theta_W\Big)\Big(\gamma_\mu\cdot\frac{1-\gamma_5}{2}\Big) -\frac{i}{3}g_1\delta_{ab}\delta_{ij}\sin\theta_W\Big(\gamma_\mu\cdot\frac{1+\gamma_5}{2}\Big) \quad (A.28)$$

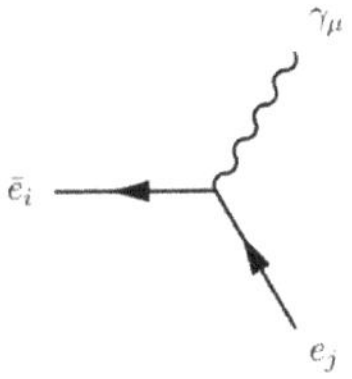

$\bar{e}_i$, e_j, γ_μ

$$\frac{i}{2}\delta_{ij}\Big(g_1\cos\theta_W+g_2\sin\theta_W\Big)\Big(\gamma_\mu\cdot\frac{1-\gamma_5}{2}\Big) +ig_1\cos\theta_W\delta_{ij}\Big(\gamma_\mu\cdot\frac{1+\gamma_5}{2}\Big) \quad (A.29)$$

$\bar{\nu}_i$, e_j, W^+_μ

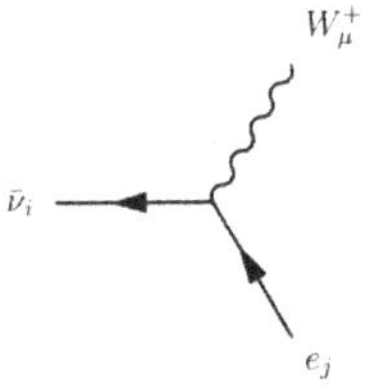

$$-i\frac{1}{\sqrt{2}}g_2V^{e,*}_{L,ji}\Big(\gamma_\mu\cdot\frac{1-\gamma_5}{2}\Big) \quad (A.30)$$

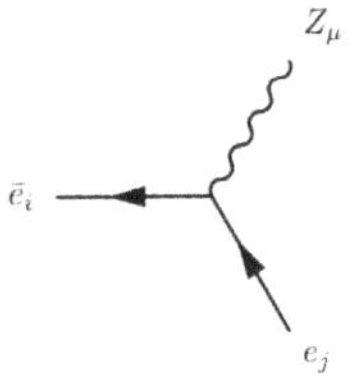

$\bar{e}_i$, e_j, Z_μ

$$\frac{i}{2}\delta_{ij}\Big(-g_1\sin\theta_W+g_2\cos\theta_W\Big)\Big(\gamma_\mu\cdot\frac{1-\gamma_5}{2}\Big) -ig_1\delta_{ij}\sin\theta_W\Big(\gamma_\mu\cdot\frac{1+\gamma_5}{2}\Big) \quad (A.31)$$

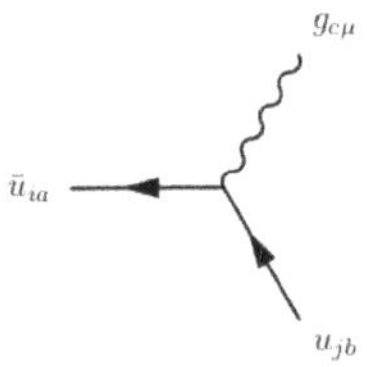

$\bar{u}_{ia}$, u_{jb}, $g_{c\mu}$

$$-\frac{i}{2}g_3\delta_{ij}\lambda^c_{a,b}\Big(\gamma_\mu\cdot\frac{1-\gamma_5}{2}\Big) -\frac{i}{2}g_3\delta_{ij}\lambda^c_{a,b}\Big(\gamma_\mu\cdot\frac{1+\gamma_5}{2}\Big) \quad (A.32)$$

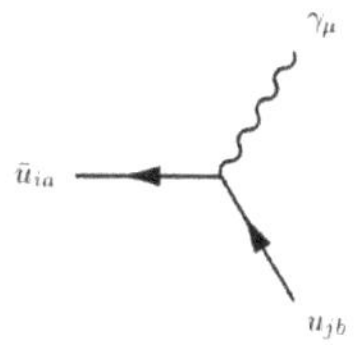

$\bar{u}_{ia}$, u_{jb}, γ_μ

$$-\frac{i}{6}\delta_{ab}\delta_{ij}\Big(3g_2\sin\theta_W + g_1\cos\theta_W\Big)\Big(\gamma_\mu\cdot\frac{1-\gamma_5}{2}\Big)$$
$$-\frac{2i}{3}g_1\cos\theta_W\delta_{ab}\delta_{ij}\Big(\gamma_\mu\cdot\frac{1+\gamma_5}{2}\Big) \qquad \text{(A.33)}$$

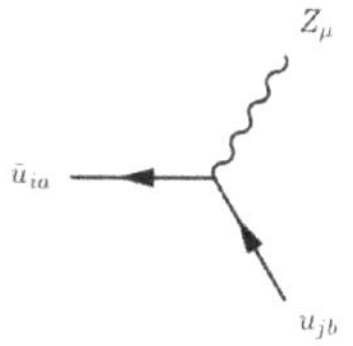

$\bar{u}_{ia}$, u_{jb}, Z_μ

$$-\frac{i}{6}\delta_{ab}\delta_{ij}\Big(3g_2\cos\theta_W - g_1\sin\theta_W\Big)\Big(\gamma_\mu\cdot\frac{1-\gamma_5}{2}\Big)$$
$$+\frac{2i}{3}g_1\delta_{ab}\delta_{ij}\sin\theta_W\Big(\gamma_\mu\cdot\frac{1+\gamma_5}{2}\Big) \qquad \text{(A.34)}$$

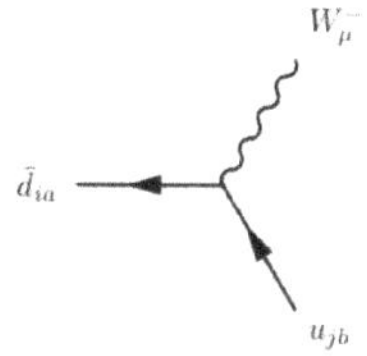

$\bar{d}_{ia}$, u_{jb}, W^-_μ

$$-i\frac{1}{\sqrt{2}}g_2\delta_{ab}\sum_{a=1}^{3}V^{u,*}_{L,ja}V^{d}_{L,ia}\Big(\gamma_\mu\cdot\frac{1-\gamma_5}{2}\Big) \qquad \text{(A.35)}$$

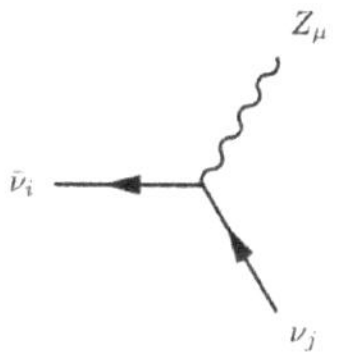

$\bar{\nu}_i$, ν_j, Z_μ

$$-\frac{i}{2}\delta_{ij}\Big(g_1\sin\theta_W + g_2\cos\theta_W\Big)\Big(\gamma_\mu\cdot\frac{1-\gamma_5}{2}\Big) \qquad \text{(A.36)}$$

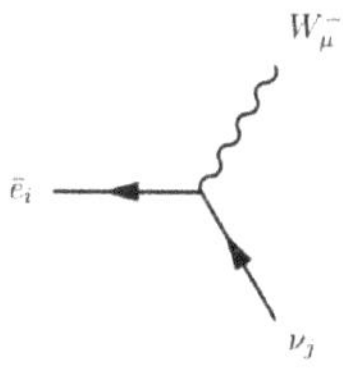

$\bar{e}_i$, ν_j, W^-_μ

$$-i\frac{1}{\sqrt{2}}g_2V^{e}_{L,ij}\Big(\gamma_\mu\cdot\frac{1-\gamma_5}{2}\Big) \qquad \text{(A.37)}$$

$\bar{d}_{ia}$, d_{jb}, G^0

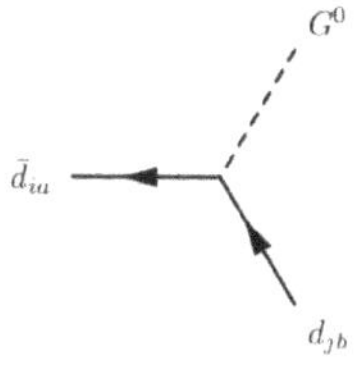

$$-\frac{1}{\sqrt{2}}\delta_{ab}\sum_{b=1}^{3}V^{d,*}_{L,jb}\sum_{a=1}^{3}V^{d,*}_{R,ia}Y_{d,ab}\Big(\frac{1-\gamma_5}{2}\Big)$$
$$+\frac{1}{\sqrt{2}}\delta_{ab}\sum_{b=1}^{3}\sum_{a=1}^{3}Y^{*}_{d,ab}V^{d}_{R,ja}V^{d}_{L,ib}\Big(\frac{1+\gamma_5}{2}\Big) \qquad \text{(A.38)}$$

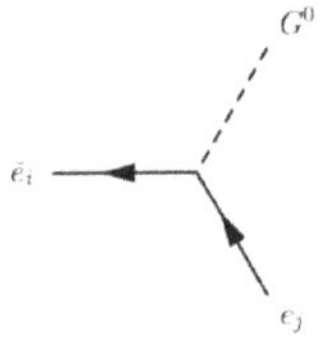

G^0, $\bar{e}_i$, e_j

$$\frac{1}{\sqrt{2}}\delta_{ab}\sum_{b=1}^{3}V_{L,jb}^{u,*}\sum_{a=1}^{3}V_{R,ia}^{u,*}Y_{u,ab}\left(\frac{1-\gamma_5}{2}\right) -\frac{1}{\sqrt{2}}\delta_{ab}\sum_{b=1}^{3}\sum_{a=1}^{3}Y_{u,ab}^{*}V_{R,ja}^{u}V_{L,ib}^{u}\left(\frac{1+\gamma_5}{2}\right) \tag{A.39}$$

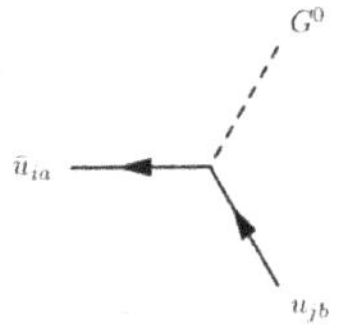

G^0, $\bar{u}_{ia}$, u_{jb}

$$-i\frac{1}{\sqrt{2}}\delta_{ab}\sum_{b=1}^{3}V_{L,jb}^{d,*}\sum_{a=1}^{3}V_{R,ia}^{d,*}Y_{d,ab}\left(\frac{1-\gamma_5}{2}\right) -i\frac{1}{\sqrt{2}}\delta_{ab}\sum_{b=1}^{3}\sum_{a=1}^{3}Y_{d,ab}^{*}V_{R,ja}^{d}V_{L,ib}^{d}\left(\frac{1+\gamma_5}{2}\right) \tag{A.40}$$

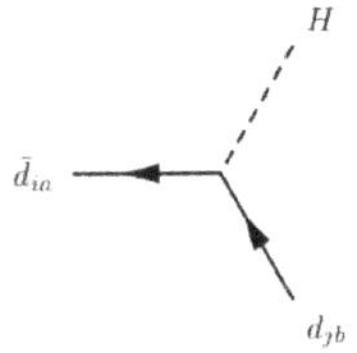

H, $\bar{d}_{ia}$, d_{jb}

$$-i\frac{1}{\sqrt{2}}\delta_{ab}\sum_{b=1}^{3}V_{L,jb}^{d,*}\sum_{a=1}^{3}V_{R,ia}^{d,*}Y_{d,ab}\left(\frac{1-\gamma_5}{2}\right) -i\frac{1}{\sqrt{2}}\delta_{ab}\sum_{b=1}^{3}\sum_{a=1}^{3}Y_{d,ab}^{*}V_{R,ja}^{d}V_{L,ib}^{d}\left(\frac{1+\gamma_5}{2}\right) \tag{A.41}$$

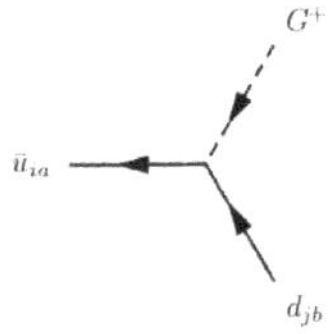

G^+, $\bar{u}_{ia}$, d_{jb}

$$i\delta_{ab}\sum_{b=1}^{3}V_{L,jb}^{d,*}\sum_{a=1}^{3}V_{R,ia}^{u,*}Y_{u,ab}\left(\frac{1-\gamma_5}{2}\right) -i\delta_{ab}\sum_{b=1}^{3}\sum_{a=1}^{3}Y_{d,ab}^{*}V_{R,ja}^{d}V_{L,ib}^{u}\left(\frac{1+\gamma_5}{2}\right) \tag{A.42}$$

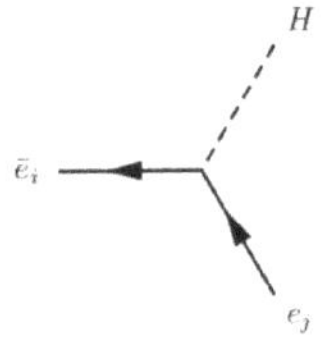

H, $\bar{e}_i$, e_j

$$-i\frac{1}{\sqrt{2}}\sum_{b=1}^{3}U_{L,jb}^{e,*}\sum_{a=1}^{3}V_{R,ia}^{e,*}Y_{e,ab}\left(\frac{1-\gamma_5}{2}\right) -i\frac{1}{\sqrt{2}}\sum_{b=1}^{3}\sum_{a=1}^{3}Y_{e,ab}^{*}V_{R,ja}^{e}V_{L,ib}^{e}\left(\frac{1+\gamma_5}{2}\right) \tag{A.43}$$

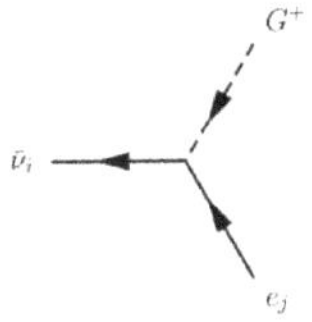

G^+, $\bar{\nu}_i$, e_j

$$-i\sum_{a=1}^{3}Y_{e,ai}^{*}V_{R,ja}^{e}\left(\frac{1+\gamma_5}{2}\right) \tag{A.44}$$

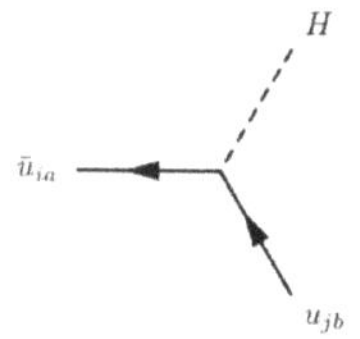

$\bar{u}_{ia}$, u_{jb}, H

$$-i\frac{1}{\sqrt{2}}\delta_{ab}\sum_{b=1}^{3}V_{L,jb}^{u,*}\sum_{a=1}^{3}V_{R,ia}^{u,*}Y_{u,ab}\left(\frac{1-\gamma_5}{2}\right)$$
$$-i\frac{1}{\sqrt{2}}\delta_{ab}\sum_{b=1}^{3}\sum_{a=1}^{3}Y_{u,ab}^{*}V_{R,ja}^{u}V_{L,ib}^{u}\left(\frac{1+\gamma_5}{2}\right) \quad (A.45)$$

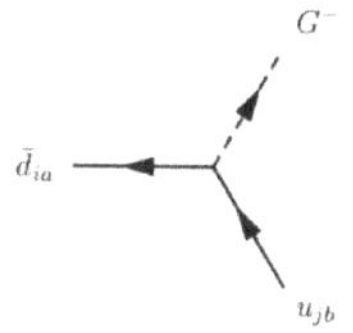

$\bar{d}_{ia}$, u_{jb}, G^-

$$-i\delta_{ab}\sum_{b=1}^{3}V_{L,jb}^{u,*}\sum_{a=1}^{3}V_{R,ia}^{d,*}Y_{d,ab}\left(\frac{1-\gamma_5}{2}\right)$$
$$+i\delta_{ab}\sum_{b=1}^{3}\sum_{a=1}^{3}Y_{u,ab}^{*}V_{R,ja}^{u}V_{L,ib}^{d}\left(\frac{1+\gamma_5}{2}\right) \quad (A.46)$$

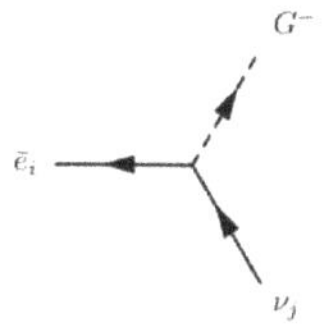

$\bar{e}_i$, ν_j, G^-

$$-i\sum_{a=1}^{3}V_{R,ia}^{e,*}Y_{e,aj}\left(\frac{1-\gamma_5}{2}\right) \quad (A.47)$$

$g_{a\rho}$, $g_{b\mu}$, $g_{c\sigma}$

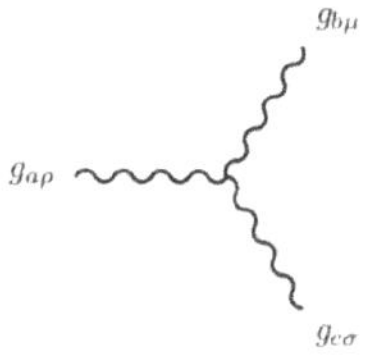

$$g_3 f_{a,b,c}\Big(g_{\rho\mu}\Big(-p_\sigma^{g_{c\mu}}+p_\sigma^{g_{a\rho}}\Big)+g_{\rho\sigma}\Big(-p_\mu^{g_{a\rho}}+p_\mu^{g_{b\sigma}}\Big)$$
$$+g_{\sigma\mu}\Big(-p_\rho^{g_{b\sigma}}+p_\rho^{g_{c\mu}}\Big)\Big) \quad (A.48)$$

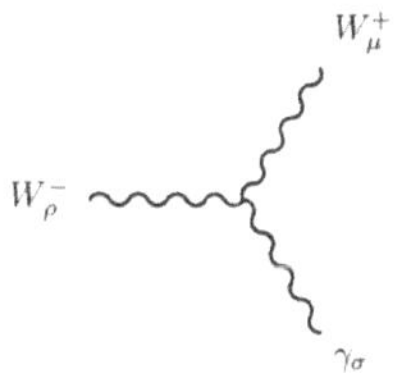

W_ρ^-, W_μ^+, γ_σ

$$-ig_2\sin\theta_W\Big(g_{\rho\mu}\Big(-p_\sigma^{W_\mu^+}+p_\sigma^{W_\rho^-}\Big)+g_{\rho\sigma}\Big(-p_\mu^{W_\rho^-}+p_\mu^{\gamma_\sigma}\Big)$$
$$+g_{\sigma\mu}\Big(-p_\rho^{\gamma_\sigma}+p_\rho^{W_\mu^+}\Big)\Big) \quad (A.49)$$

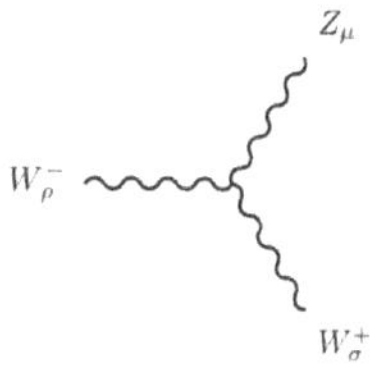

W_ρ^-, Z_μ, W_σ^+

$$ig_2\cos\theta_W\Big(g_{\rho\mu}\Big(-p_\sigma^{Z_\mu}+p_\sigma^{W_\rho^-}\Big)+g_{\rho\sigma}\Big(-p_\mu^{W_\rho^-}+p_\mu^{W_\sigma^+}\Big)$$
$$+g_{\sigma\mu}\Big(-p_\rho^{W_\sigma^+}+p_\rho^{Z_\mu}\Big)\Big) \quad (A.50)$$

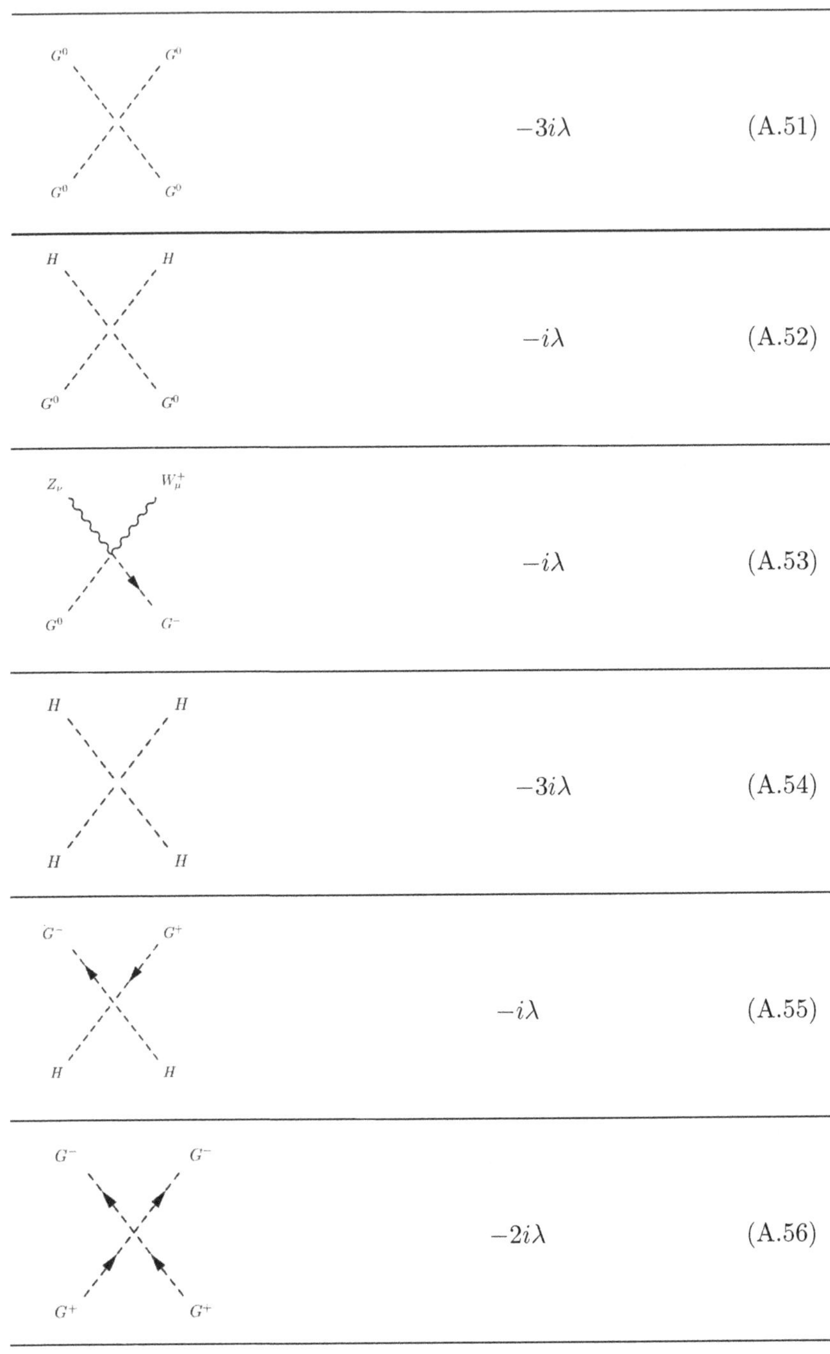

$-3i\lambda$	(A.51)
$-i\lambda$	(A.52)
$-i\lambda$	(A.53)
$-3i\lambda$	(A.54)
$-i\lambda$	(A.55)
$-2i\lambda$	(A.56)

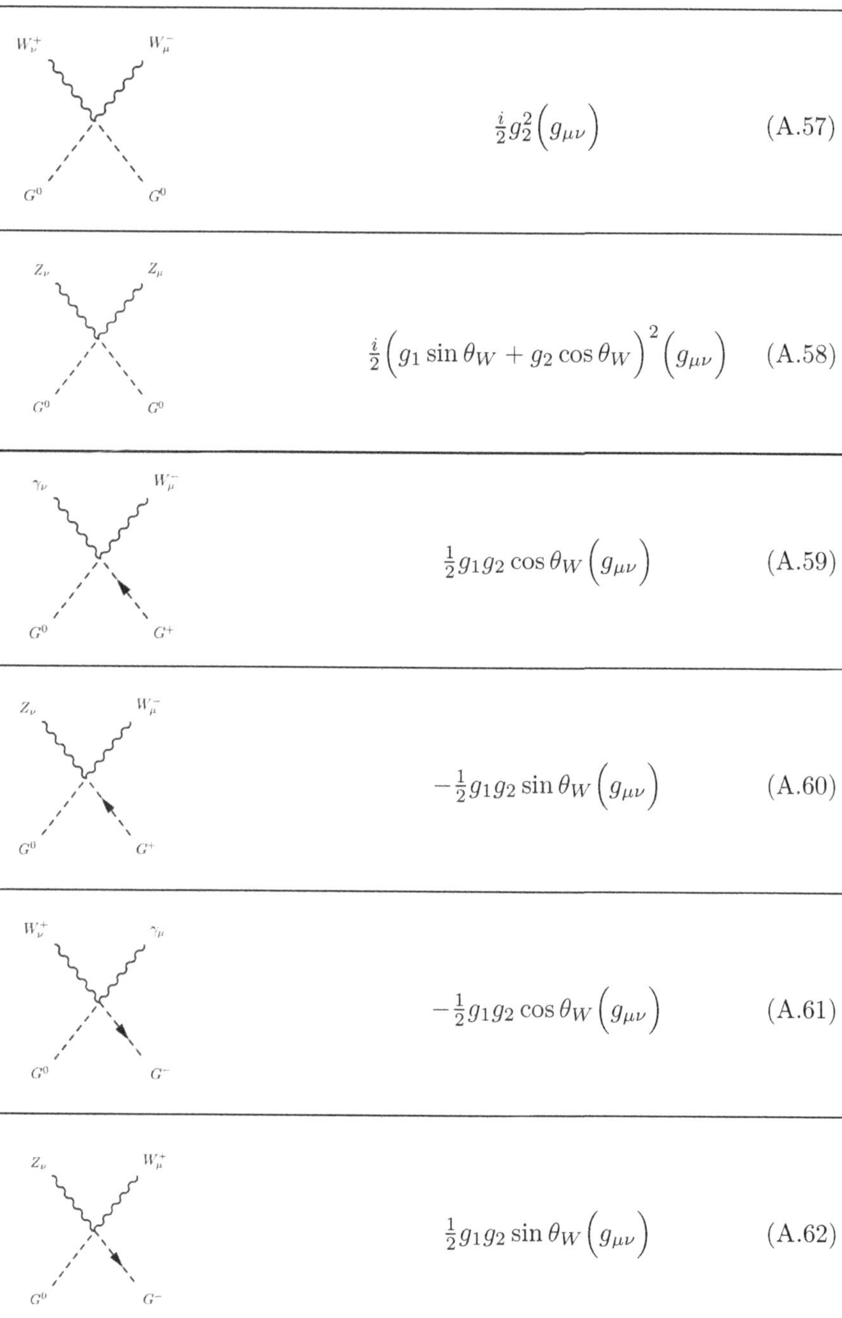

$$\frac{i}{2} g_2^2 \Big(g_{\mu\nu} \Big) \tag{A.57}$$

$$\frac{i}{2} \Big(g_1 \sin\theta_W + g_2 \cos\theta_W \Big)^2 \Big(g_{\mu\nu} \Big) \tag{A.58}$$

$$\frac{1}{2} g_1 g_2 \cos\theta_W \Big(g_{\mu\nu} \Big) \tag{A.59}$$

$$-\frac{1}{2} g_1 g_2 \sin\theta_W \Big(g_{\mu\nu} \Big) \tag{A.60}$$

$$-\frac{1}{2} g_1 g_2 \cos\theta_W \Big(g_{\mu\nu} \Big) \tag{A.61}$$

$$\frac{1}{2} g_1 g_2 \sin\theta_W \Big(g_{\mu\nu} \Big) \tag{A.62}$$

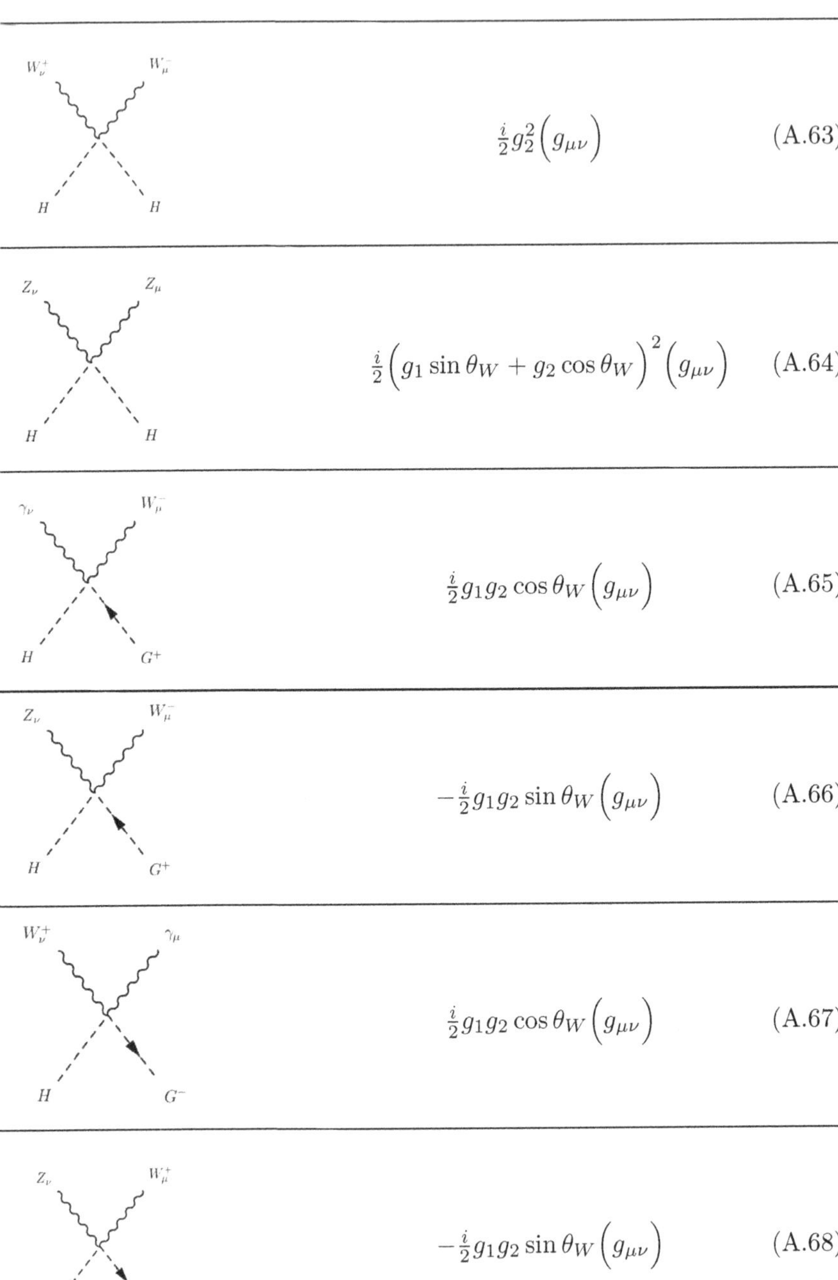

$$\frac{i}{2} g_2^2 \Big(g_{\mu\nu} \Big) \tag{A.63}$$

$$\frac{i}{2} \Big(g_1 \sin\theta_W + g_2 \cos\theta_W \Big)^2 \Big(g_{\mu\nu} \Big) \tag{A.64}$$

$$\frac{i}{2} g_1 g_2 \cos\theta_W \Big(g_{\mu\nu} \Big) \tag{A.65}$$

$$-\frac{i}{2} g_1 g_2 \sin\theta_W \Big(g_{\mu\nu} \Big) \tag{A.66}$$

$$\frac{i}{2} g_1 g_2 \cos\theta_W \Big(g_{\mu\nu} \Big) \tag{A.67}$$

$$-\frac{i}{2} g_1 g_2 \sin\theta_W \Big(g_{\mu\nu} \Big) \tag{A.68}$$

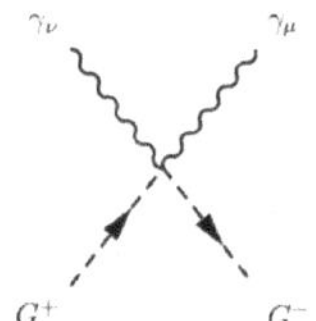

$$\tfrac{i}{2}\Big(g_1\cos\theta_W + g_2\sin\theta_W\Big)^2\Big(g_{\mu\nu}\Big) \quad \text{(A.69)}$$

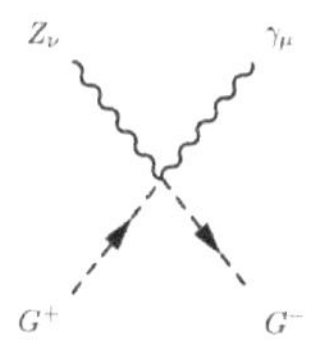

$$\tfrac{i}{4}\Big(2g_1g_2\cos 2\theta_W + \Big(g_2^2 - g_1^2\Big)\sin 2\theta_W\Big)g_{\mu\nu} \quad \text{(A.70)}$$

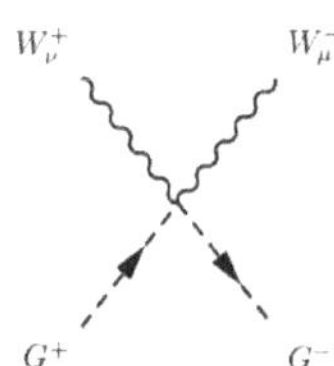

$$\tfrac{i}{2}g_2^2\Big(g_{\mu\nu}\Big) \quad \text{(A.71)}$$

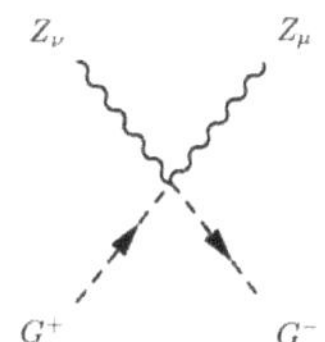

$$\tfrac{i}{2}\Big(-g_1\sin\theta_W + g_2\cos\theta_W\Big)^2\Big(g_{\mu\nu}\Big) \quad \text{(A.72)}$$

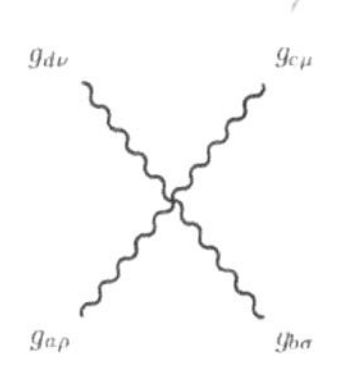

$$\begin{aligned}
&-ig_3^2\Big(\textstyle\sum_{a=1}^{8} f_{a,d,a}f_{b,c,a} + \sum_{a=1}^{8} f_{a,c,a}f_{b,d,a}\Big)g_{\rho\sigma}g_{\mu\nu}\\
&+ig_3^2\Big(-\textstyle\sum_{a=1}^{8} f_{a,b,a}f_{c,d,a} + \sum_{a=1}^{8} f_{a,d,a}f_{b,c,a}\Big)g_{\rho\mu}g_{\sigma\nu}\\
&+ig_3^2\Big(\textstyle\sum_{a=1}^{8} f_{a,c,a}f_{b,d,a} + \sum_{a=1}^{8} f_{a,b,a}f_{c,d,a}\Big)g_{\rho\nu}g_{\sigma\mu} \quad \text{(A.73)}
\end{aligned}$$

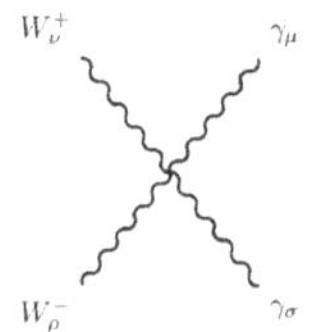

$$\begin{aligned}
&ig_2^2\sin\theta_W^2\Big(g_{\rho\sigma}g_{\mu\nu}\Big)\\
&+ig_2^2\sin\theta_W^2\Big(g_{\rho\mu}g_{\sigma\nu}\Big)\\
&-2ig_2^2\sin\theta_W^2\Big(g_{\rho\nu}g_{\sigma\mu}\Big) \quad \text{(A.74)}
\end{aligned}$$

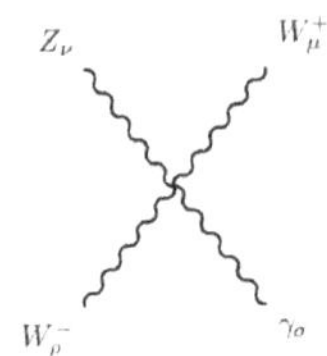

$$\begin{aligned} &ig_2^2 \cos\theta_W \sin\theta_W \left(g_{\rho\sigma} g_{\mu\nu}\right) \\ &-ig_2^2 \sin 2\theta_W \left(g_{\rho\mu} g_{\sigma\nu}\right) \\ &+ig_2^2 \cos\theta_W \sin\theta_W \left(g_{\rho\nu} g_{\sigma\mu}\right) \end{aligned} \tag{A.75}$$

W_ν^+ W_μ^+ W_ρ^- W_σ^-

$$\begin{aligned} &2ig_2^2 \left(g_{\rho\sigma} g_{\mu\nu}\right) - ig_2^2 \left(g_{\rho\mu} g_{\sigma\nu}\right) \\ &-ig_2^2 \left(g_{\rho\nu} g_{\sigma\mu}\right) \end{aligned} \tag{A.76}$$

Z_ν Z_μ W_ρ^- W_σ^+

$$\begin{aligned} &-2ig_2^2 \cos\theta_W^2 \left(g_{\rho\sigma} g_{\mu\nu}\right) + ig_2^2 \cos\theta_W^2 \left(g_{\rho\mu} g_{\sigma\nu}\right) \\ &+ig_2^2 \cos\theta_W^2 \left(g_{\rho\nu} g_{\sigma\mu}\right) \end{aligned} \tag{A.77}$$

$g_{c\mu}$ $\bar{\eta}_a^G$ η_b^G

$$g_3 f_{a,b,c} \left(p_\mu^{\eta_b^G}\right) \tag{A.78}$$

W_μ^+ $\bar{\eta}^+$ η^γ

$$-ig_2 \sin\theta_W \left(p_\mu^{\eta^\gamma}\right) \tag{A.79}$$

W_μ^- $\bar{\eta}^-$ η^γ

$$ig_2 \sin\theta_W \left(p_\mu^{\eta^\gamma}\right) \tag{A.80}$$

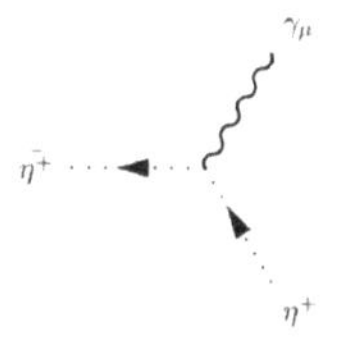

$$ig_2 \sin\theta_W \left(p_\mu^{\eta^+}\right) \tag{A.81}$$

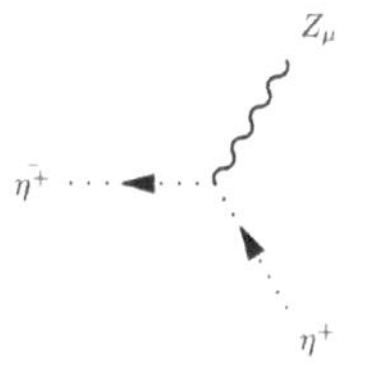

$$ig_2 \cos\theta_W \left(p_\mu^{\eta^+}\right) \tag{A.82}$$

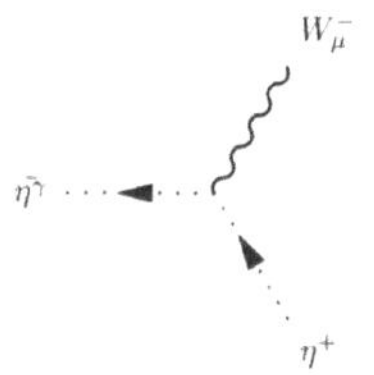

$$-ig_2 \sin\theta_W \left(p_\mu^{\eta^+}\right) \tag{A.83}$$

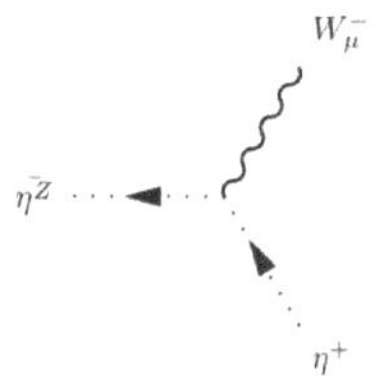

$$-ig_2 \cos\theta_W \left(p_\mu^{\eta^+}\right) \tag{A.84}$$

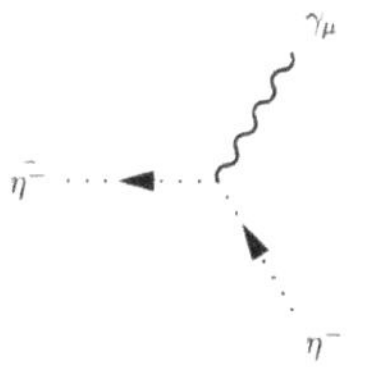

$$-ig_2 \sin\theta_W \left(p_\mu^{\eta^-}\right) \tag{A.85}$$

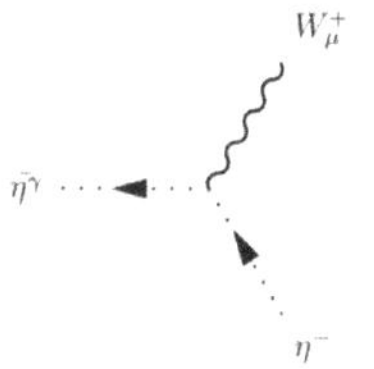

$$ig_2 \sin\theta_W \left(p_\mu^{\eta^-}\right) \tag{A.86}$$

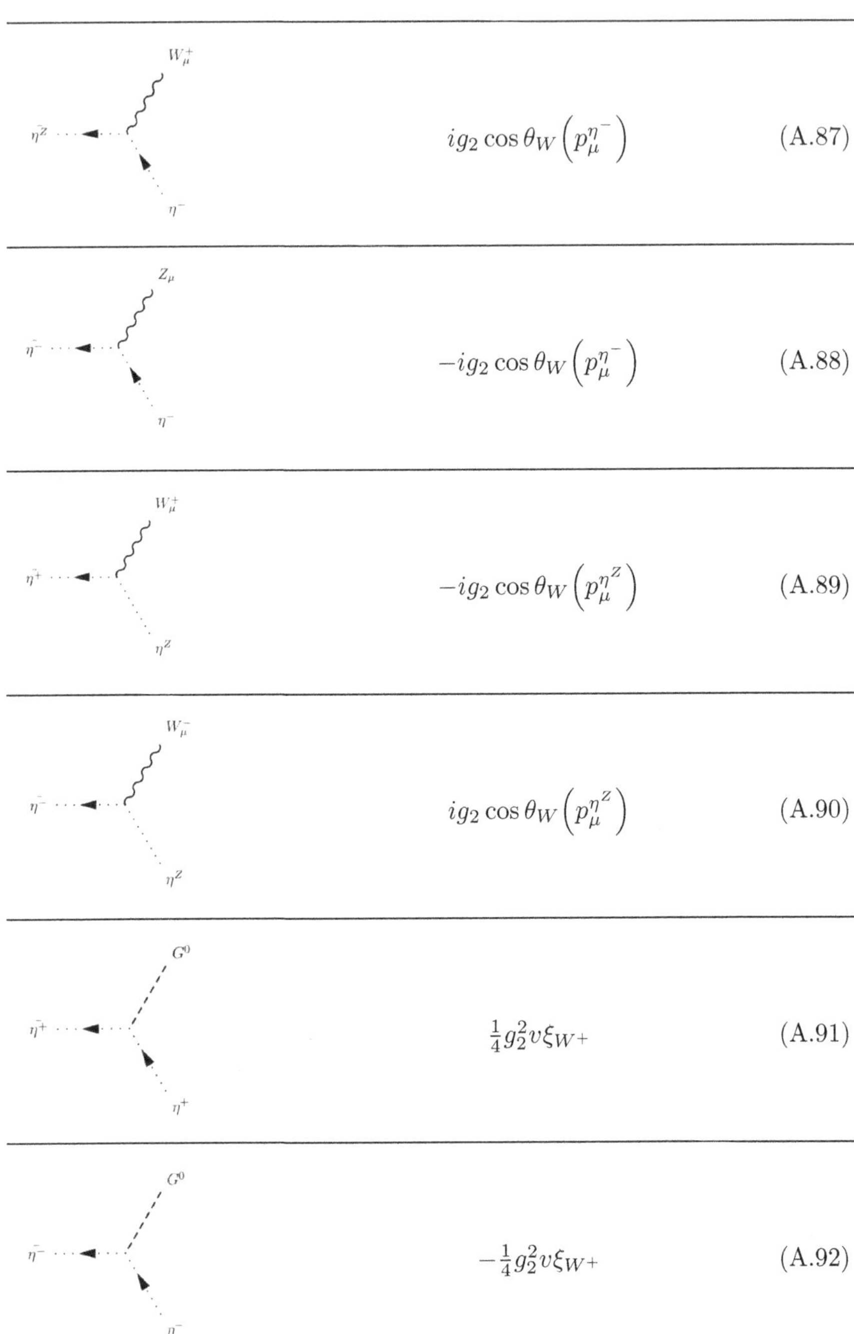

W_μ^+
$\bar{\eta}^Z$
η^-
$ig_2 \cos\theta_W \left(p_\mu^{\eta^-}\right)$
(A.87)
Z_μ
$\bar{\eta}^-$
η^-
$-ig_2 \cos\theta_W \left(p_\mu^{\eta^-}\right)$
(A.88)
W_μ^+
$\bar{\eta}^+$
η^Z
$-ig_2 \cos\theta_W \left(p_\mu^{\eta^Z}\right)$
(A.89)
W_μ^-
$\bar{\eta}^-$
η^Z
$ig_2 \cos\theta_W \left(p_\mu^{\eta^Z}\right)$
(A.90)
G^0
$\bar{\eta}^+$
η^+
$\frac{1}{4} g_2^2 v \xi_{W^+}$
(A.91)
G^0
$\bar{\eta}^-$
η^-
$-\frac{1}{4} g_2^2 v \xi_{W^+}$
(A.92)

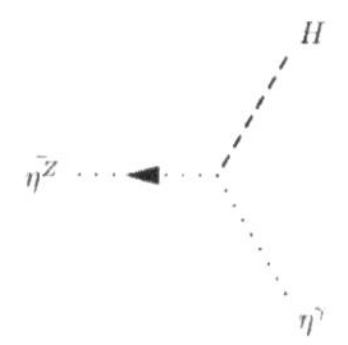

$$\frac{i}{8}v\xi_Z\Big(2g_1g_2\cos 2\theta_W+\Big(-g_2^2+g_1^2\Big)\sin 2\theta_W\Big) \tag{A.93}$$

$$-\frac{i}{4}g_2v\xi_{W^+}\Big(g_1\cos\theta_W+g_2\sin\theta_W\Big) \tag{A.94}$$

$$-\frac{i}{4}g_2v\xi_{W^+}\Big(g_1\cos\theta_W+g_2\sin\theta_W\Big) \tag{A.95}$$

$$-\frac{i}{4}g_2^2v\xi_{W^+} \tag{A.96}$$

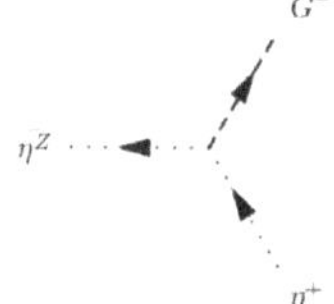

$$\frac{i}{4}g_2v\xi_Z\Big(g_1\sin\theta_W+g_2\cos\theta_W\Big) \tag{A.97}$$

$$-\frac{i}{4}g_2^2v\xi_{W^+} \tag{A.98}$$

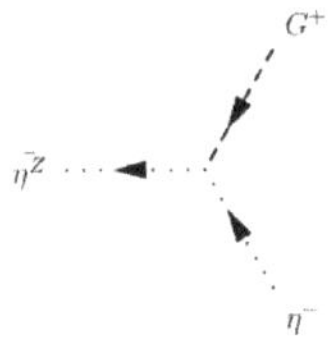

Diagram	Vertex factor
$\bar{\eta}^Z$, G^+, η^-	$\frac{i}{4}g_2 v\xi_Z\Big(g_1\sin\theta_W + g_2\cos\theta_W\Big)$ (A.99)
$\bar{\eta}^Z$, H, η^Z	$-\frac{i}{4}v\xi_Z\Big(g_1\sin\theta_W + g_2\cos\theta_W\Big)^2$ (A.100)
$\bar{\eta}^+$, G^+, η^Z	$-\frac{i}{4}g_2 v\xi_{W^+}\Big(-g_1\sin\theta_W + g_2\cos\theta_W\Big)$ (A.101)
$\bar{\eta}^-$, G^-, η^Z	$-\frac{i}{4}g_2 v\xi_{W^+}\Big(-g_1\sin\theta_W + g_2\cos\theta_W\Big)$ (A.102)

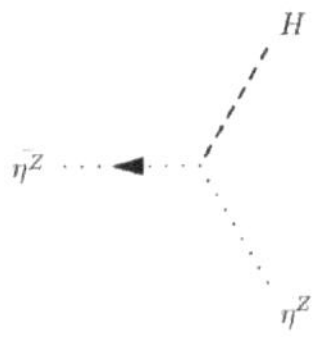

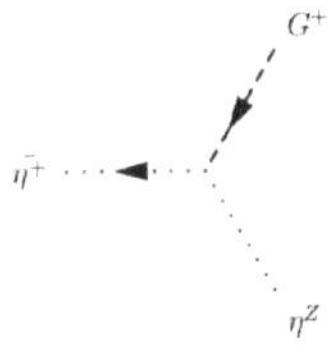

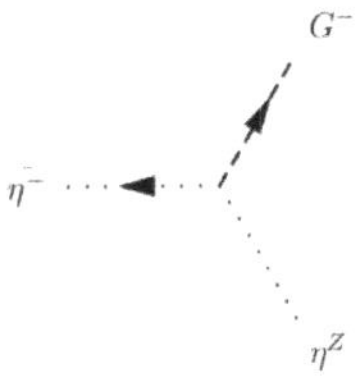

Bibliography

[1] I. Kant. *The Critique of Pure Reason (translated by J.M.D. Meiklejohn)*. CreateSpace Independent Publishing Platform, 2011.

[2] D. Hanneke, S. Fogwell Hoogerheide, and G. Gabrielse. Cavity Control of a Single-Electron Quantum Cyclotron: Measuring the Electron Magnetic Moment. *Phys. Rev. A*, 83:052122, 2011.

[3] T. Aoyama, M. Hayakawa, T. Kinoshita, and M. Nio. Tenth-order QED Contribution to the Electron $g - 2$ and an Improved Value of the Fine Structure Constant. *Phys. Rev. Lett.*, 109:111807, 2012.

[4] T. Aoyama, M. Hayakawa, T. Kinoshita, and M. Nio. Tenth-order Electron Anomalous Magnetic Moment: Contribution of Diagrams without Closed Lepton Loops. *Phys. Rev. D*, 91(3):033006, 2015. [Erratum: Phys. Rev. D 96, 019901 (2017)].

[5] C.-N. Yang. Fermi's β-decay Theory. *Asia Pac. Phys. Newslett.*, 01:27–30, 2012.

[6] F. Joliot and I. Curie. Un nouveau type de radioactivité. *J. Phys. (in French)*, 5 (153):254, 1934.

[7] C. S. Wu, E. Ambler, R. W. Hayward, D. D. Hoppes, and R. P. Hudson. Experimental Test of Parity Conservation in β Decay. *Phys. Rev.*, 105:1413–1414, 1957.

[8] T. D. Lee and C. Yang. Question of Parity Conservation in Weak Interactions. *Phys. Rev.*, 104:254–258, 1956.

[9] R. L. Garwin, L. M. Lederman, and M. Weinrich. Observations of the Failure of Conservation of Parity and Charge Conjugation in Meson Decays: The Magnetic Moment of the Free Muon. *Phys. Rev.*, 105:1415–1417, 1957.

[10] L. D. Landau. On the Conservation Laws for Weak Interactions. *Nucl. Phys.*, 3:127–131, 1957.

[11] A. Salam. On Parity Conservation and Neutrino Mass. *Nuovo Cim.*, 5:299–301, 1957.

[12] T. D. Lee and C.-N. Yang. Parity Nonconservation and a Two Component Theory of the Neutrino. *Phys. Rev.*, 105:1671–1675, 1957.

[13] R. P. Feynman and M. Gell-Mann. Theory of Fermi Interaction. *Phys. Rev.*, 109:193–198, 1958.

[14] E. C. G. Sudarshan and R. E. Marshak. Chirality Invariance and the Universal Fermi Interaction. *Phys. Rev.*, 109:1860–1860, 1958.

[15] C. L. Cowan, F. Reines, F. B. Harrison, H. W. Kruse, and A. D. McGuire. Detection of the Free Neutrino: A Confirmation. *Science*, 124:103–104, 1956.

[16] N. Cabibbo. Unitary Symmetry and Leptonic Decays. *Phys. Rev. Lett.*, 10:531–533, 1963.

[17] J. D. Bjorken and S. L. Glashow. Elementary Particles and $SU(4)$. *Phys. Lett.*, 11:255–257, 1964.

[18] S. L. Glashow, J. Iliopoulos, and L. Maiani. Weak Interactions with Lepton-Hadron Symmetry. *Phys. Rev. D*, 2:1285–1292, 1970.

[19] J. J. Aubert et al. Experimental Observation of a Heavy Particle J. *Phys. Rev. Lett.*, 33:1404–1406, 1974.

[20] J. E. Augustin et al. Discovery of a Narrow Resonance in e^+e^- Annihilation. *Phys. Rev. Lett.*, 33:1406–1408, 1974. [Adv. Exp. Phys. 5, 141 (1976)].

[21] M. L. Perl. The Discovery of the τ Lepton. In 3rd *International Symposium on the History of Particle Physics: The Rise of the Standard Model*, pages 79–100, 9 1992.

[22] S. W. Herb et al. Observation of a Dimuon Resonance at 9.5-GeV in 400-GeV Proton-Nucleus Collisions. *Phys. Rev. Lett.*, 39:252–255, 1977.

[23] F. Abe et al. Observation of Top Quark Production in $\bar{p}p$ Collisions. *Phys. Rev. Lett.*, 74:2626–2631, 1995.

[24] S. Abachi et al. Observation of the Top Quark. *Phys. Rev. Lett.*, 74:2632–2637, 1995.

[25] J. Horejsi. *Introduction to Electroweak Unification: Standard Model from Tree Unitarity.* 6 1993.

[26] C. Yang and R. L. Mills. Conservation of Isotopic Spin and Isotopic Gauge Invariance. *Phys. Rev.*, 96:191–195, 1954.

[27] S. L. Glashow. Partial Symmetries of Weak Interactions. *Nucl. Phys.*, 22:579–588, 1961.

[28] K. Nishijima. Charge Independence Theory of V Particles. *Prog. Theor. Phys.*, 13(3):285–304, 1955.

[29] M. Gell-Mann. The interpretation of the New Particles As Displaced Charge Multiplets. *Nuovo Cim.*, 4(S2):848–866, 1956.

[30] S. Weinberg. A Model of Leptons. *Phys. Rev. Lett.*, 19:1264–1266, 1967.

[31] A. Salam. Elementary Particle Physics (Nobel Symp. N.8). *Stockholm*, page 367, 1968.

[32] J. Goldstone. Field Theories with Superconductor Solutions. *Nuovo Cim.*, 19:154–164, 1961.

[33] P. W. Anderson. Plasmons, Gauge Invariance, and Mass. *Phys. Rev.*, 130:439–442, 1963.

[34] F. Englert and R. Brout. Broken Symmetry and the Mass of Gauge Vector Mesons. *Phys. Rev. Lett.*, 13:321–323, 1964.

[35] P. W. Higgs. Broken Symmetries and the Masses of Gauge Bosons. *Phys. Rev. Lett.*, 13:508–509, 1964.

[36] G. S. Guralnik, C. R. Hagen, and T. W. B. Kibble. Global Conservation Laws and Massless Particles. *Phys. Rev. Lett.*, 13:585–587, 1964.

[37] G. 't Hooft. Renormalization of Massless Yang-Mills Fields. *Nucl. Phys. B*, 33:173–199, 1971.

[38] G. 't Hooft. Renormalizable Lagrangians for Massive Yang-Mills Fields. *Nucl. Phys. B*, 35:167–188, 1971.

[39] G. 't Hooft and M. J. G. Veltman. Regularization and Renormalization of Gauge Fields. *Nucl. Phys.*, B44:189–213, 1972.

[40] P. Z. Skands. QCD for Collider Physics. In *Proceedings, High-energy Physics. Proceedings, 18th European School (ESHEP 2010): Raseborg, Finland, June 20 – July 3, 2010*, 2011.

[41] R. K. Ellis, W. J. Stirling, and B. R. Webber. QCD and Collider Physics. *Camb. Monogr. Part. Phys. Nucl. Phys. Cosmol.*, 8:1–435, 1996.

[42] M. E. Peskin and D. V. Schroeder. *An Introduction to Quantum Field Theory.* Addison-Wesley, Reading, USA, 1995.

[43] D. J. Gross and F. Wilczek. Ultraviolet Behavior of Nonabelian Gauge Theories. *Phys. Rev. Lett.*, 30:1343–1346, 1973.

[44] H. D. Politzer. Reliable Perturbative Results for Strong Interactions? *Phys. Rev. Lett.*, 30:1346–1349, 1973.

[45] S. Bethke. World Summary of α_s (2012). 2012. [Nucl. Phys. Proc. Suppl. 234, 229 (2013)].

[46] S. Bethke. The 2009 World Average of α_s. *Eur. Phys. J. C*, 64:689–703, 2009.

[47] T. Regge. Introduction to Complex Orbital Momenta. *Nuovo Cim.*, 14:951, 1959.

[48] B. Andersson, G. Gustafson, G. Ingelman, and T. Sjostrand. Parton Fragmentation and String Dynamics. *Phys. Rept.*, 97:31–145, 1983.

[49] B. R. Webber. A QCD Model for Jet Fragmentation Including Soft Gluon Interference. *Nucl. Phys. B*, 238:492–528, 1984.

[50] G. Marchesini and B. R. Webber. Simulation of QCD Jets Including Soft Gluon Interference. *Nucl. Phys. B*, 238:1–29, 1984.

[51] G. Marchesini and B. R. Webber. Monte Carlo Simulation of General Hard Processes with Coherent QCD Radiation. *Nucl. Phys. B*, 310:461–526, 1988.

[52] J. Ellis. The Discovery of the Gluon. *Int. J. Mod. Phys.*, A29(31):1430072, 2014.

[53] M. Gell-Mann. The Eightfold Way: A Theory of Strong Interaction Symmetry. 3 1961.

[54] Y. Ne'eman. Derivation of Strong Interactions From a Gauge Invariance. *Nucl. Phys.*, 26:222–229, 1961.

[55] M. Gell-Mann. A Schematic Model of Baryons and Mesons. *Phys. Lett.*, 8:214–215, 1964.

[56] G. Zweig. An $SU(3)$ Model for Strong Interaction Symmetry and its Breaking. Version 1. 1964.

[57] G. Zweig. An $SU(3)$ Model for Strong Interaction Symmetry and its Breaking. Version 2. In D.B. Lichtenberg and Simon Peter Rosen, editors, *Developments in the Quark Theory of Hadrons. VOL. 1. 1964 - 1978*, pages 22–101. 1964.

[58] O. W. Greenberg. Spin and Unitary Spin Independence in a Paraquark Model of Baryons and Mesons. *Phys. Rev. Lett.*, 13:598–602, 1964.

[59] M. Y. Han and Y. Nambu. Three Triplet Model with Double $SU(3)$ Symmetry. *Phys. Rev.*, 139:B1006–B1010, 1965.

[60] E. D. Bloom et al. High-Energy Inelastic *ep* Scattering at 6-Degrees and 10-Degrees. *Phys. Rev. Lett.*, 23:930–934, 1969.

[61] J. D. Bjorken. Applications of the Chiral $U(6) \times U(6)$ Algebra of Current Densities. *Phys. Rev.*, 148:1467–1478, 1966.

[62] J. D. Bjorken and E. A. Paschos. Inelastic Electron Proton and gamma Proton Scattering, and the Structure of the Nucleon. *Phys. Rev.*, 185:1975–1982, 1969.

[63] J. D. Bjorken. Theoretical Ideas on Inelastic Electron and Muon Scattering. In *Proceedings: 3rd International Symposium on Electron and Photon Interactions at High Energies, SLAC: Stanford, USA, September 5–9, 1967*, pages 109–127, 1967.

[64] R. P. Feynman. Very High-Energy Collisions of Hadrons. *Phys. Rev. Lett.*, 23:1415–1417, 1969.

[65] J. C. G. Callan and D. J. Gross. High-Energy Electroproduction and the Constitution of the Electric Current. *Phys. Rev. Lett.*, 22:156–159, 1969.

[66] C. H. Llewellyn Smith. Inelastic Lepton Scattering in Gluon Models. *Phys. Rev.*, D4:2392, 1971.

[67] B. R. Stella and H.-J. Meyer. Y(9.46 GeV) and the Gluon Discovery (a Critical Recollection of PLUTO Results). *Eur. Phys. J. H*, 36:203–243, 2011.

[68] C. Berger et al. Jet Analysis of the Υ (9.46) Decay Into Charged Hadrons. *Phys. Lett. B*, 82:449–455, 1979.

[69] C. Berger et al. Topology of the Υ Decay. *Z. Phys. C*, 8:101, 1981.

[70] R. Brandelik et al. Evidence for Planar Events in e^+ e^- Annihilation at High-Energies. *Phys. Lett. B*, 86:243–249, 1979.

[71] D. P. Barber et al. Discovery of Three Jet Events and a Test of Quantum Chromodynamics at PETRA Energies. *Phys. Rev. Lett.*, 43:830, 1979.

[72] C. Berger et al. Evidence for Gluon Bremsstrahlung in e^+ e^- Annihilations at High-Energies. *Phys. Lett. B*, 86:418–425, 1979.

[73] W. Bartel et al. Observation of Planar Three Jet Events in e^+ e^- Annihilation and Evidence for Gluon Bremsstrahlung. *Phys. Lett. B*, 91:142–147, 1980.

[74] R. Brandelik et al. Evidence for a Spin One Gluon in Three Jet Events. *Phys. Lett. B*, 97:453–458, 1980.

[75] C. Berger et al. A Study of Multi-Jet Events in e^+ e^- Annihilation. *Phys. Lett. B*, 97:459–464, 1980.

[76] P. Soding. On the Discovery of the Gluon. *Eur. Phys. J. H*, 35:3–28, 2010.

[77] R. M. Barnett et al. Review of Particle Physics. Particle Data Group. *Phys. Rev.*, D54:1–720, 1996.

[78] G. F. Sterman and S. Weinberg. Jets From Quantum Chromodynamics. *Phys. Rev. Lett.*, 39:1436, 1977.

[79] G. Wolf et al. Tasso Results on e^+e^- Annihilation Between 13-GeV and 31.6-GeV and Evidence for Three Jet Events. *eConf*, C790823:34, 1979.

[80] S. Moretti and J. B. Tausk. QCD Effects and b-Tagging at LEP-1. *Z. Phys.*, C69:635–646, 1996.

[81] Z. Kunszt, P. Nason, G. Marchesini, and B. R. Webber. QCD at LEP. In *LEP Physics Workshop Geneva, Switzerland, February 20, 1989*, pages 373–453, 1989.

[82] S. Kluth. Jet Physics in e^+e^- Annihilation From 14-GeV to 209-GeV. *Nucl. Phys. Proc. Suppl.*, 133:36–46, 2004.

[83] Z. Kunszt, S. Moretti, and W. J. Stirling. Higgs Production at the LHC: An Update on Cross-sections and Branching Ratios. *Z. Phys.*, C74:479–491, 1997.

[84] J. F. Gunion, H. E. Haber, G. L. Kane, and S. Dawson. The Higgs Hunter's Guide. *Front. Phys.*, 80:1–404, 2000.

[85] J. F. Gunion, H. E. Haber, G. L. Kane, and S. Dawson. Errata for the Higgs Hunter's Guide. 1992.

[86] B. A. Kniehl. Higgs Phenomenology at One Loop in the Standard Model. *Phys. Rept.*, 240:211–300, 1994.

[87] M. Spira. QCD Effects in Higgs Physics. *Fortsch. Phys.*, 46:203–284, 1998.

[88] A. Djouadi. The Anatomy of Electro-weak Symmetry Breaking. I: The Higgs Boson in the Standard Model. *Phys. Rept.*, 457:1–216, 2008.

[89] W. Keung and W. J. Marciano. Higgs Scalar Decays : $H \to W^{\pm}X$. *Phys. Rev.*, D30:248, 1984.

[90] R. N. Cahn. The Higgs Boson. *Rept. Prog. Phys.*, 52:389, 1989.

[91] T. Asaka and K. Hikasa. Four Fermion Decay of Higgs Bosons. *Phys. Lett.*, B345:36–41, 1995.

[92] A. Barroso, J. Pulido, and J. C. Romao. Higgs Production at e^+e^- Colliders. *Nucl. Phys.*, B267:509–530, 1986.

[93] A. Grau, G. Panchieri, and R. J. N. Phillips. Contributions of Off-Shell Top Quarks to Decay processes. *Phys. Lett.*, B251:293–298, 1990.

[94] S. Moretti and W. J. Stirling. Contributions of Below Threshold Decays to MSSM Higgs Branching Ratios. *Phys. Lett.*, B347:291–299, 1995. [Erratum: Phys. Lett. B366, 451 (1996)].

[95] S. G. Gorishny, A. L. Kataev, S. A. Larin, and L. R. Surguladze. Corrected Three Loop QCD Correction to the Correlator of the Quark Scalar Currents and $\Gamma_{Tot}(H^0 \to \text{Hadrons})$. *Mod. Phys. Lett. A*, 5:2703–2712, 1990.

[96] L. R. Surguladze. Quark Mass Effects in Fermionic Decays of the Higgs Boson in $\mathcal{O}(\alpha_s^2)$ Perturbative QCD. *Phys. Lett.*, B341:60–72, 1994.

[97] T. Inami, T. Kubota, and Y. Okada. Effective Gauge Theory and the Effect of Heavy Quarks in Higgs Boson Decays. *Z. Phys.*, C18:69–80, 1983.

[98] S. Dawson and R. P. Kauffman. QCD Corrections to $H \to \gamma\gamma$. *Phys. Rev.*, D47:1264–1267, 1993.

[99] M. Drees and K. Hikasa. Note on QCD Corrections to Hadronic Higgs Decay. *Phys. Lett.*, B240:455, 1990. [Erratum: Phys. Lett. B262, 497 (1991)].

[100] A. Djouadi, M. Spira, and P. M. Zerwas. Two Photon Decay Widths of Higgs Particles. *Phys. Lett.*, B311:255–260, 1993.

[101] Z. Kunszt and W. J. Stirling. The Standard Model Higgs at LHC: Branching Ratios and Cross Sections. In *ECFA Large Hadron Collider Workshop, Aachen, Germany, 4–9 Oct 1990: Proceedings.2.*, pages 428–443, 1991.

[102] M. Spira, A. Djouadi, D. Graudenz, and P. M. Zerwas. Higgs Boson Production at the LHC. *Nucl. Phys.*, B453:17–82, 1995.

[103] H. M. Georgi, S. L. Glashow, M. E. Machacek, and D. V. Nanopoulos. Higgs Bosons From Two Gluon Annihilation in Proton-Proton Collisions. *Phys. Rev. Lett.*, 40:692, 1978.

[104] R. N. Cahn and S. Dawson. Production of Very Massive Higgs Bosons. *Phys. Lett.*, 136B:196, 1984. [Erratum: Phys. Lett. 138B, 464 (1984)].

[105] S. L. Glashow, D. V. Nanopoulos, and A. Yildiz. Associated Production of Higgs Bosons and Z Particles. *Phys. Rev.*, D18:1724–1727, 1978.

[106] Z. Kunszt, Z. Trocsanyi, and W. J. Stirling. Clear Signal of Intermediate Mass Higgs Boson Production at LHC and SSC. *Phys. Lett.*, B271:247–255, 1991.

[107] Z. Kunszt. Associated Production of Heavy Higgs Boson with Top Quarks. *Nucl. Phys.*, B247:339–359, 1984.

[108] J. F. Gunion, H. E. Haber, F. E. Paige, W. Tung, and S. S. D. Willenbrock. Neutral and Charged Higgs Detection: Heavy Quark Fusion, Top Quark Mass Dependence and Rare Decays. *Nucl. Phys.*, B294:621, 1987.

[109] D. Graudenz, M. Spira, and P. M. Zerwas. QCD Corrections to Higgs Boson Production at Proton-Proton Colliders. *Phys. Rev. Lett.*, 70:1372–1375, 1993.

[110] M. Spira, A. Djouadi, and P. M. Zerwas. QCD Corrections to the $HZ\gamma$ Coupling. *Phys. Lett.*, B276:350–353, 1992.

[111] S. Dawson. Radiative Corrections to Higgs Boson Production. *Nucl. Phys. B*, 359:283–300, 1991.

[112] S. Dawson and R. Kauffman. QCD Corrections to Higgs Boson Production: Nonleading Terms in the Heavy Quark Limit. *Phys. Rev.*, D49:2298–2309, 1994.

[113] A. Djouadi. Higgs Particles at Future Hadron and Electron-Positron Colliders. *Int. J. Mod. Phys.*, A10:1–64, 1995.

[114] S. Dittmaier et al. Handbook of LHC Higgs Cross Sections: 1. Inclusive Observables. 1 2011.

[115] T. Han, G. Valencia, and S. Willenbrock. Structure Function Approach to Vector Boson Scattering in pp Collisions. *Phys. Rev. Lett.*, 69:3274–3277, 1992.

[116] W. Beenakker, S. Dittmaier, M. Kramer, B. Plumper, M. Spira, and P. M. Zerwas. Higgs Radiation off Top Quarks at the Tevatron and the LHC. *Phys. Rev. Lett.*, 87:201805, 2001.

[117] S. Dawson, C. Jackson, L. H. Orr, L. Reina, and D. Wackeroth. Associated Higgs Production with Top Quarks at the Large Hadron Collider: NLO QCD Corrections. *Phys. Rev. D*, 68:034022, 2003.

[118] M. Kobayashi and T. Maskawa. CP Violation in the Renormalizable Theory of Weak Interaction. *Prog. Theor. Phys.*, 49:652–657, 1973.

[119] Y. Nambu. Axial Vector Current Conservation in Weak Interactions. *Phys. Rev. Lett.*, 4:380–382, 1960.

[120] K. Hagiwara et al. Review of Particle Physics. Particle Data Group. *Phys. Rev. D*, 66:010001, 2002.

[121] L. Wolfenstein. Parametrization of the Kobayashi-Maskawa Matrix. *Phys. Rev. Lett.*, 51:1945, 1983.

[122] M. Bona et al. The UTfit Collaboration Report on The Status of The Unitarity Triangle Beyond the Standard Model. I. Model-independent Analysis and Minimal Flavor Violation. *JHEP*, 03:080, 2006.

[123] C. Jarlskog. Commutator of the Quark Mass Matrices in the Standard Electroweak Model and a Measure of Maximal CP Violation. *Phys. Rev. Lett.*, 55:1039, 1985.

[124] C. Jarlskog. A Basis Independent Formulation of the Connection Between Quark Mass Matrices, CP Violation and Experiment. *Z. Phys.*, C29:491–497, 1985.

[125] S. Herrlich and U. Nierste. The Complete $|\Delta S| = 2$ Hamiltonian in the Next-to-leading Order. *Nucl. Phys. B*, 476:27–88, 1996.

[126] T. Inami and C. S. Lim. Effects of Superheavy Quarks and Leptons in Low-Energy Weak Processes $K(L) \to \mu^+\mu^-$, $K^+ \to \pi + \nu\bar{\nu}$ and

$K^0 \leftrightarrow \bar{K}^0$. *Prog. Theor. Phys.*, 65:297, 1981. [Erratum: Prog. Theor. Phys. 65, 1772 (1981)].

[127] M. E. Gamiz Sanchez. Kaon Physics: CP Violation and Hadronic Matrix Elements. Other thesis, 10 2003.

[128] Y. Nir. CP violation: A New era. In *Heavy Flavor Physics: Theory and Experimental Results in Heavy Quark Physics and CP Violation. Proceedings, 55th Scottish Universities Summer School in Physics, SUSSP 2001, St. Andrews, UK, August 7–23, 2001*, pages 147–200, 2001.

[129] J. R. Batley et al. A Precision Measurement of Direct CP Violation in the Decay of Neutral Kaons into Two Pions. *Phys. Lett. B*, 544:97–112, 2002.

[130] G. D. Barr et al. A New Measurement of Direct CP Violation in the Neutral Kaon System. *Phys. Lett. B*, 317:233–242, 1993.

[131] E. Abouzaid et al. Precise Measurements of Direct CP Violation, CPT Symmetry, and Other Parameters in the Neutral Kaon System. *Phys. Rev. D*, 83:092001, 2011.

[132] L. K. Gibbons et al. Measurement of the CP Violation Parameter Re(ϵ'/ϵ). *Phys. Rev. Lett.*, 70:1203–1206, 1993.

[133] F. J. Gilman and M. B. Wise. Effective Hamiltonian for $\Delta s = 1$ Weak Nonleptonic Decays in the Six Quark Model. *Phys. Rev. D*, 20:2392, 1979.

[134] A. J. Buras. Weak Hamiltonian, CP Violation and Rare Decays. In *Les Houches Summer School in Theoretical Physics, Session 68: Probing the Standard Model of Particle Interactions*, pages 281–539, 6 1998.

[135] A. J. Buras, M. Jamin, and M. E. Lautenbacher. The Anatomy of ϵ'/ϵ Beyond Leading Logarithms with Improved Hadronic Matrix Elements. *Nucl. Phys. B*, 408:209–285, 1993.

[136] A. J. Buras, P. Gambino, M. Gorbahn, S. Jager, and L. Silvestrini. ϵ'/ϵ and Rare K and B Decays in the MSSM. *Nucl. Phys. B*, 592:55–91, 2001.

[137] A. Alavi-Harati et al. Measurements of Direct CP Violation, CPT Symmetry, and Other Parameters in the Neutral Kaon System. *Phys. Rev. D*, 67:012005, 2003. [Erratum: Phys. Rev. D 70, 079904 (2004)].

[138] E. Abouzaid et al. Precise Measurements of Direct CP Violation, CPT Symmetry, and Other Parameters in the Neutral Kaon System. *Phys. Rev. D*, 83:092001, 2011.

[139] A. J. Buras. Flavor Dynamics: CP Violation and Rare Decays. *Subnucl. Ser.*, 38:200–337, 2002.

[140] A. Abulencia et al. Observation of $B_s^0 - \bar{B}_s^0$ Oscillations. *Phys. Rev. Lett.*, 97:242003, 2006.

[141] V. M. Abazov et al. First Direct Two-sided Bound on the B_s^0 Oscillation Frequency. *Phys. Rev. Lett.*, 97:021802, 2006.

[142] C. H. Llewellyn Smith. High-Energy Behavior and Gauge Symmetry. *Phys. Lett. B*, 46:233–236, 1973.

[143] J. M. Cornwall, D. N. Levin, and G. Tiktopoulos. Uniqueness of Spontaneously Broken Gauge Theories. *Phys. Rev. Lett.*, 30:1268–1270, 1973. [Erratum: Phys. Rev. Lett. 31, 572 (1973)].

[144] J. M. Cornwall, D. N. Levin, and G. Tiktopoulos. Derivation of Gauge Invariance From High-Energy Unitarity Bounds on the S Matrix. *Phys. Rev. D*, 10:1145, 1974. [Erratum: Phys. Rev. D 11, 972 (1975)].

[145] S. D. Joglekar. S Matrix Derivation of the Weinberg Model. *Ann. Phys.*, 83:427, 1974.

[146] S. Weinberg. Physical Processes in a Convergent Theory of the Weak and Electromagnetic Interactions. *Phys. Rev. Lett.*, 27:1688–1691, 1971.

[147] J. Schechter and Y. Ueda. High-energy Behavior of Gauge-theory Tree Graphs. (erratum). *Phys. Rev.*, D7:3119–3123, 1973. [Erratum: Phys. Rev. D8, 3709 (1973)].

[148] J. Schechter and Y. Ueda. Higgs Mesons and High-energy Behavior. *Lett. Nuovo Cim.*, 8S2:991–993, 1973. [Lett. Nuovo Cim. 8, 991 (1973)].

[149] M. A. Furman and G. J. Komen. Scaling Behavior of Charged Spin 1 Partons in a Gauge Model. *Nucl. Phys.*, B84:323–332, 1975.

[150] B. W. Lee, C. Quigg, and H. B. Thacker. Weak Interactions at Very High-Energies: The Role of the Higgs Boson Mass. *Phys. Rev.*, D16:1519, 1977.

[151] B. W. Lee, C. Quigg, and H. B. Thacker. The Strength of Weak Interactions at Very High-Energies and the Higgs Boson Mass. *Phys. Rev. Lett.*, 38:883–885, 1977.

[152] F. J. Hasert et al. Search for Elastic ν_μ Electron Scattering. *Phys. Lett.*, 46B:121–124, 1973.

[153] F. J. Hasert et al. Observation of Neutrino Like Interactions Without Muon Or Electron in the Gargamelle Neutrino Experiment. *Phys. Lett.*, 46B:138–140, 1973.

[154] S. Schael et al. Precision Electroweak Measurements on the Z Resonance. *Phys. Rept.*, 427:257–454, 2006.

[155] P. A. Zyla et al. Review of Particle Physics. *PTEP*, 2020(8):083C01, 2020.

[156] F. Abe et al. Evidence for Top Quark Production in $\bar{p}p$ collisions at $\sqrt{s} = 1.8$ TeV. *Phys. Rev. Lett.*, 73:225–231, 1994.

[157] R. Barate et al. Search for the Standard Model Higgs boson at LEP. *Phys. Lett. B*, 565:61–75, 2003.

[158] S. Eidelman. Gauge & Higgs Boson Summary Table. *Phys. Lett.*, B592:31–88, 2004.

[159] The Tevatron New Physics Higgs Working Group. Combined CDF and D0 Upper Limits on Standard Model Higgs-Boson Production with up to 6.7 fb^{-1} of Data. In *Proceedings, 35th International Conference on High Energy Physics (ICHEP 2010): Paris, France, July 22–28, 2010*, 2010.

[160] G. Aad et al. Observation of a New Particle in the Search for the Standard Model Higgs Boson with the ATLAS Detector at the LHC. *Phys. Lett.*, B716:1–29, 2012.

[161] S. Chatrchyan et al. Observation of a New Boson at a Mass of 125 GeV with the CMS Experiment at the LHC. *Phys. Lett.*, B716:30–61, 2012.

[162] The Tevatron New Physics Higgs Working Group. Updated Combination of CDF and D0 Searches for Standard Model Higgs Boson Production with up to 10.0 fb^{-1} of Data. 2012.

[163] ATLAS Collaboration. A Combination of Measurements of Higgs Boson Production and Decay Using up to 139 fb^{-1} of Proton-Proton Collision Data at $\sqrt{s} = 13$ TeV Collected with the ATLAS Experiment. 2020.

[164] CMS Collaboration. Measurement of Higgs Boson Decay to a Pair of Muons in Proton-Proton Collisions at $\sqrt{s} = 13\,\mathrm{TeV}$. 2020.

[165] R. Davis D. S. Harmer, and K. C. Hoffman. Search for Neutrinos From the Sun. *Phys. Rev. Lett.*, 20:1205–1209, 1968.

[166] Y. Fukuda et al. Evidence for Oscillation of Atmospheric Neutrinos. *Phys. Rev. Lett.*, 81:1562–1567, 1998.

[167] Q. R. Ahmad et al. Measurement of the Rate of $\nu_e + d \to p + p + e^-$ Interactions Produced by ^{8}B Solar Neutrinos at the Sudbury Neutrino Observatory. *Phys. Rev. Lett.*, 87:071301, 2001.

[168] B. Pontecorvo. Neutrino Experiments and the Problem of Conservation of Leptonic Charge. *Sov. Phys. JETP*, 26:984–988, 1968. [Zh. Eksp. Teor. Fiz. 53, 1717 (1967)].

[169] Z. Maki, M. Nakagawa, and S. Sakata. Remarks on the Unified Model of Elementary Particles. *Prog. Theor. Phys.*, 28:870–880, 1962.

[170] S. Eidelman et al. Review of Particle Physics. Particle Data Group. *Phys. Lett.*, B592(1–4):1, 2004.

[171] Y. Abe et al. Indication of Reactor $\bar{\nu}_e$ Disappearance in the Double Chooz Experiment. *Phys. Rev. Lett.*, 108:131801, 2012.

[172] T. Yanagida. Horizontal Symmetry and Masses of Neutrinos. *Prog. Theor. Phys.*, 64:1103, 1980.

[173] R. N. Mohapatra and G. Senjanovic. Neutrino Mass and Spontaneous Parity Nonconservation. *Phys. Rev. Lett.*, 44:912, 1980.

[174] M. Magg and C. Wetterich. Neutrino Mass Problem and Gauge Hierarchy. *Phys. Lett.*, 94B:61–64, 1980.

[175] R. N. Mohapatra and G. Senjanovic. Neutrino Masses and Mixings in Gauge Models with Spontaneous Parity Violation. *Phys. Rev.*, D23:165, 1981.

[176] T. P. Cheng and L. Li. Neutrino Masses, Mixings and Oscillations in $SU(2)\times U(1)$ Models of Electroweak Interactions. *Phys. Rev.*, D22:2860, 1980.

[177] E. Ma. Pathways to Naturally Small Neutrino Masses. *Phys. Rev. Lett.*, 81:1171–1174, 1998.

[178] R. Foot, H. Lew, X. G. He, and Girish C. Joshi. Seesaw Neutrino Masses Induced by a Triplet of Leptons. *Z. Phys. C*, 44:441, 1989.

[179] R. N. Mohapatra and J. W. F. Valle. Neutrino Mass and Baryon Number Nonconservation in Superstring Models. *Phys. Rev.*, D34:1642, 1986.

[180] G. 't Hooft. Naturalness, Chiral Symmetry, and Spontaneous Chiral Symmetry Breaking. *NATO Sci. Ser. B*, 59:135–157, 1980.

[181] J. Bauer, M. Gartmeier, H. Gruber, and H. Heid. *Vocations and Learning*, 1:87, 2008.

[182] S. Khalil and S. Moretti. *Supersymmetry Beyond Minimality: From Theory to Experiment.* CRC Press (Taylor & Frances), December 2017.

[183] Florian Staub. SARAH 4: A Tool for (not only SUSY) Model Builders. *Comput. Phys. Commun.*, 185:1773–1790, 2014.

Subject index

Printed in the United States
by Baker & Taylor Publisher Services